THE SCIENTIST'S GUIDE TO WRITING

THE SCIENTIST'S GUIDE TO WRITING

HOW TO WRITE MORE EASILY AND EFFECTIVELY THROUGHOUT YOUR SCIENTIFIC CAREER

STEPHEN B. HEARD

PRINCETON UNIVERSITY PRESS

PRINCETON AND OXFORD

press.princeton.edu

Cover image courtesy of Shutterstock.

Author's note: The cover image shows a stylized image of Z-DNA. The usual conformation of DNA is a right-handed helix, with major and minor grooves. Under certain conditions, however, DNA reshapes itself into the left-handed helix shown here. This can happen even in cells, and be involved in regulating gene expression. Science is rich and fascinating.

ISBN 978-0-691-17021-3

ISBN (pbk.) 978-0-691-17022-0

Library of Congress Control Number: 2015950552

British Library Cataloging-in-Publication Data is available

An earlier version of chapter 28 has been published as "On whimsy, jokes, and beauty: can scientific writing be enjoyed?"
(SB Heard, 2014, Ideas in Ecology and Evolution 7:65-72).

This book has been composed in Minion Pro

Printed on acid-free paper. ∞

Printed in the United States of America

5 7 9 10 8 6 4

Contents

Preface   vii

Part I. What Writing Is

1. On Bacon, Hobbes, and Newton, and the
   Selfishness of Writing Well   3

2. Genius, Craft, and What This Book Is About   11

Part II. Behavior

3. Reading   17

4. Managing Your Writing Behavior   22

5. Getting Started   30

6. Momentum   42

Part III. Content and Structure

7. Finding and Telling Your Story   57

8. The Canonical Structure of the Scientific Paper   74

9. Front Matter and Abstract   79

10. The Introduction Section   84

11. The Methods Section   89

12. The Results Section   99

13. The Discussion Section   120

14. Back Matter   126

15. Citations   132

16. Deviations from the IMRaD Canon   138

Part IV. Style

17. Paragraphs   149

18. Sentences   159

19. Words   174

20. Brevity   182

Part V. Revision

21. Self-Revision   193

22. Friendly Review   204

23. Formal Review   211

24. Revision and the "Response to Reviews"   222

Part VI. Some Loose Threads

25. The Diversity of Writing Forms   233

26. Managing Coauthorships   247

27. Writing in English for Non-Native Speakers   260

Part VII. Final Thoughts

28. On Whimsy, Jokes, and Beauty: Can Scientific Writing Be Enjoyed?   273

Acknowledgements   287

References   289

Permanent URLs   299

Index   303

That you're dipping into this book probably means that you suspect writing is important to your scientific career, and that you'd like to do it better, or more quickly, or more easily.

You're right—writing *is* important to a career in science: at least as important as disciplinary knowledge, or experimental design, or statistics. As a scientist, you'll work hard to make new discoveries about the world; but only writing (and publication) makes what you've learned part of human knowledge. Not only that: only by writing (and publishing) can you advance in your career—get cited, get fellowships, get hired, get promoted. Because it's so important, you may spend more time writing than you do designing or executing experiments; over your career, you'll probably produce a startlingly large amount of writing. A typical year for me involves something like seventy-five thousand words (almost the length of this book), and I've kept that up, year after year, for about twenty-five years now. Your own pace is unlikely to be much slower.

If you aren't entirely satisfied with the way you write now, you're not alone. Writing is hard for everyone—the new writer and the seasoned veteran. Fortunately, writing is a craft, and as you practice the craft, you can improve. Not only that, but you can improve faster if you pay deliberate attention to practicing and learning. This book offers help. I'll give you perspective on why scientists write as we do, guidance about your goals as a scientific writer, and advice about how to manage yourself as a writer to reach those goals. Along the way, I'll ask you to think about the *structure*, *content*, and *style* of what you write, and also about the *process* by which you write it—that is, about your behavior and psychology as a writer.

This book is designed for students and early-career scientists across the natural sciences (including mathematics). I'm a biologist, an evolutionary ecologist in particular, so you might wonder how I know which writing advice is good for cell biologists, physicists, earth scientists, or pure mathematicians. Well, a great deal of what a scientific writer needs to know is universal: we all face the same behavioral challenges in getting writing done, we all want to write so our work will be understood and cited, we all use a common set of tools (elements of English composition, graphic design, and so on), and we all take our writing through the same review and publication process. There are differences among fields, of course: for example, conventions for order of authorship vary, and writers in pure mathematics construct Introductions and Discussions rather differently from those in other fields. In order to discover differences such as these, I've read hundreds of scientific papers across the breadth of the sciences and talked with friends and colleagues who work in fields different from mine. I'm sure there are field-specific writing details that I don't cover, but these will be minor compared with the common aims we all hold and the shared techniques we all use as scientific writers.

Because I'm a scientist, I like to present writing advice as I do anything else: with data. So ideally, my suggestions would be supported by replicated, controlled experiments—imagine hundreds of individually caged scientific writers, each assigned randomly to write in the passive or the active voice—or at least by systematic and well-replicated observational studies. Where such data exist, I cite them, but the scientific study of scientific writing is much less advanced than you might think. To fill the (many) gaps, I also offer advice that distills my own experience as a writer, reviewer, editor, and teacher of writing. Over my twenty-five years as a scientist, I've written or co-written about seventy journal papers and book chapters and many, many more grant proposals, technical reports, administrative documents, lay essays, blog posts, and so on. I've been a peer reviewer for hundreds of journal-submitted manuscripts, and I've handled hundreds more as an associate editor. Finally, I've advised dozens of graduate students (my own and my colleagues') as they've worked—and sometimes struggled—to write theses, papers, grants, and so on. From all this experience I've learned more than a few things and formed more than a few opinions, and in the following pages I'll offer you both.

Of course, you can't learn writing only by reading about it. It's important to apply what you read to your own writing and, more broadly, to practice writing as much and as deliberately as you can. For this reason, I offer exercises at the end of most chapters. They'll work best if you do them along with a colleague or classmate and share and discuss your answers. Writing *is* important to your career in science, so it's worth investing some real effort in developing your skills. No one ever completely masters our writing craft, but if you take it seriously you can get much, much better—and that pays off. Good luck!

PART I

What Writing Is

Scientists spend enormous amounts of time writing. Many spend more time on writing than on designing experiments, gathering and analyzing data, devising proofs, or any of the other things scientists do. And yet, many scientists pay little attention to writing as a process. They think of it as a rather mechanical step in which they simply record the results of the work after they're "done."

This view of writing is badly misleading. For most of us, writing is hard work, a source of stress and frustration, and so it deserves the same kind of deliberate consideration we give to experimental design. What is it that you're trying to write, and why do the standard scientific forms you use have the structures, styles, and other attributes they do? What belongs in a manuscript, what doesn't, and why? What are you actually thinking and doing as you sit at the keyboard writing (or, perhaps, not writing)? What's the relationship between the writer and the reader, and how can deliberate thought about that relationship make your writing better?

The central message of this book is that all these questions can be answered, and the quality and quantity of your writing vastly increased, with attention to three points. First, most scientific writers aren't born geniuses, but develop facility with writing by deliberately practicing the craft. Second, the goal of all scientific writing is clarity: effortless transfer of information or argument from writer to reader. Third, it's enormously helpful for writers to think consciously about their own writing behavior. This book will explore all of these points at length. We'll begin, though, with something fundamental but often overlooked: if you want to get better at writing, it helps to think about what writing is—by which I mean how we write, why we write that way, and how that "way" has evolved over the years to better suit our needs as writers and as readers.

ONE

‖‖‖

On Bacon, Hobbes, and Newton, and the Selfishness of Writing Well

The Invention of Clarity

In the European early modern period (c.1500–1750), everything was changing. The period saw the Protestant Reformation, the introduction of representative democracy, the secularization of political power, and the origins of the sovereign nation-state. It saw globalization of trade in goods and ideas, but also the subjugation of much of the world under European colonization.

Science was transforming itself right alongside religion, politics, and global economies. European curiosity cabinets (Figure 1.1) were bulging with specimens returned from overseas exploration and trade: stones, creatures, and artifacts begging to be explained by new ideas in natural science and anthropology. Chemistry took its first steps away from alchemy and toward systematic, rational discovery. Astronomy and physics were revolutionized by painstaking observations and new instruments, and by increasing openness of thought about the data these yielded. Finally, the invention and application of the calculus put mathematics at the center of all the sciences.

But while the *content* of human knowledge was exploding, another, more important change was taking place. The development of modern scientific methods, professional scientists, scientific societies, learned journals, and (in case you were wondering about the point of this historical excursion) modern-style scientific writing changed the way people acquired and communicated knowledge. In a sense, this was when scientists learned to write—or more particularly, to write with the explicit goal of making their ideas available to a broad scientific community.

Figure 1.1 Frontispiece to Ole Worm's (1655) Museum Wormianum, a catalog of his curiosity cabinet.

Writing to communicate with the scientific community was a big change. Medieval "scientists" (alchemists, for instance) generally thought of themselves as solitary workers who would penetrate nature's secrets for their own gain. Thus, if they wrote their findings down at all it was to claim priority or to make notes for their own use—and what they wrote was deliberately obscure, even written in code, cryptic symbols, or anagrams, to protect their secrets from their rivals. One of the first proponents of change was Francis Bacon, who criticized this secrecy and argued instead in his 1609 essay *De Sapientia Veterum* that "perfection of the sciences is to be looked for not from the swiftness or ability of any one inquirer, but from a succession." In the posthumously published *New Atlantis* (1627), Bacon described a fictitious research institute–cum–scientific society he called "Salomon's House"—and he clearly intended his utopian novel to be a proposal for how science should work. In Salomon's House research progressed because scientists communicated and collaborated with one another. (Bacon might well have been inspired by Islamic science of the 8th and 9th centuries, which flourished, collaboratively, under the Abbasid caliphs Harun al-Rashid and Abu al-Mamun [Lyons 2009].)

Bacon's concept of Salomon's House inspired the creation of the Royal Society of London in 1660. Its founders extended his ideas about communication among collaborating scientists to communication with a broad scientific community and even with the curious public. One of those founders was Robert Boyle, who essentially invented a new form of writing: the scientific report, which described the methods and results of an experiment (Pérez-Ramos 1996). Another was Thomas Hobbes, who wrote in the preface to his 1655 work *De Corpore*, "I distinguish the most common notions by accurate definition, for the avoiding of confusion and obscurity" (xiii)—a goal that seems routine today, but would have been outrageously unconventional in Hobbes's time. The founding of the Society brought with it the first modern scientific journal, *Philosophical Transactions of the Royal Society*, which printed scientific reports of the kind pioneered by Boyle, written in the clear language advocated by Hobbes. Just a dozen years later, Thomas Sprat described the organization's rhetorical philosophy as

> a constant resolution, to reject all the amplifications, digressions, and swellings of style . . . a close, naked, natural way of speaking; positive expressions, clear sense, a native easiness: bringing all things as near the mathematical plainness[1], as they can: and preferring the language of artisans, countrymen, and merchants, before that of wits, or scholars. (Sprat 1667, 113).

All this may seem obvious from our modern vantage point, but the transition from medieval secrecy through Bacon and Hobbes to the "clear sense [and] native easiness" of Sprat's Royal Society was revolutionary. Without this tectonic shift in how science was reported, modern science couldn't be done. The inventions of the calculus, the telescope, the microscope, and the inductive method (all between 1590 and 1630) were certainly important, but they're all outweighed in im-

---

[1] This mention of "mathematical" plainness might be a shout-out to Euclid, whose *Elements* are admirably lucid. However, clarity and openness were not necessarily the rule among ancient Greek thinkers. Pythagoras, for example, bound his followers to secrecy, and his followers may have killed the philosopher Hippasus for divulging his discovery of the irrational numbers.

portance by the idea of describing one's scientific thinking clearly, for all to read.

Of course, no revolution lacks holdouts, and the revolution in scientific communication had a curious one: the famously cranky Isaac Newton, for whom publication remained largely about ensuring credit for his work. For example, he drafted his *On Analysis by Infinite Series* in 1669 in response to Nicholas Mercator's *Logorithmotechnia*, which Newton worried would undermine his claim of first discovery for some key insights underlying the calculus. Despite pressure from colleagues, Newton allowed only limited circulation of the manuscript within the Royal Society; not until 1711 would he agree to open publication. More famously, he deliberately made his masterwork *Principia Mathematica*—and especially its third volume, *De mundi systemate*—difficult to read. Newton had originally written *De mundi systemate* in plain language to be accessible to readers (Westfall 1980, 459) but changed his mind and rewrote it as series of propositions, derivations, lemmas, and proofs comprehensible only to accomplished mathematicians. He left little doubt of his intent, telling his friend William Derham that "in order to avoid being baited by little smatterers in mathematics, he [Newton] designedly made his *Principia* abstruse" (Derham 1733). That is, he wrote to impede communication with other scientists, not to facilitate it! Of course, by then Newton was a superstar whose writing was likely to command from his readers whatever effort was needed to penetrate the fog. And readers could spare the effort, as the flow of published works competing for scientists' attention was still little more than a trickle. This, too, would change.

Clarity and "Telepathy" in the Modern Era

Bacon, Hobbes, Sprat, and others of their time were taking the first steps toward what became, by the twentieth century, a consensus that the goal of most writing is clear communication. The best-known reflection of this is probably *The Elements of Style*, by William Strunk Jr. and E. B. White, first published in 1920. White described Strunk's opinion that the typical reader was "floundering in a swamp" and that it was "the duty of anyone trying to write English to drain this swamp quickly and get [the reader] up on dry ground, or at least throw [down] a rope" (Strunk and White 1972, xii). However forceful Strunk's pleading, though, the argument for clarity has its purest expression in Stephen King's 2000 *On Writing: A Memoir of the Craft.* King's chapter "What Writing Is" opens with the simple declaration: "Telepathy, of course" (King 2000, 95).

The word "telepathy" may seem chosen for humor, but in scientific writing your goal should always be communication so crystal clear that it feels to the reader like telepathy—like direct transmission to the reader's brain from yours. You are writing because you have some information to transmit, and your goal should be for the reader to receive that information without even being aware of the process. As Nathaniel Hawthorne put it, "The greatest possible merit of style is . . . to make the words absolutely disappear into the thought" (letter to E. A. Duyckinck, 27 Apr. 1851, quoted in Van Doren [1949], 267). If the reader pauses to question your word choice or needs to squint to distinguish between two lines on a graph, then you have joined a battle you don't want to be in: what you're trying to say is fighting for the reader's attention with the way you're saying it.

At this point you might be a little skeptical. After all, popular wisdom holds that people who use big words and complicated sentences seem more intelligent. (A quick internet search will turn up dozens of lists of "Ten Words That Will Make You Sound Smarter.") There is even some limited support for this idea: one study compared undergraduates reading texts printed in easier- and harder-to-read fonts (to vary reading difficulty independent of content) and found they scored the harder-to-read texts as better written (Galak and Nelson 2011). However, most research finds the opposite: that people ascribe higher intelligence to writers who (and higher quality to texts that) use clearer fonts, smaller words, and simpler sentences (e.g., Oppenheimer 2006). Even if difficult prose did make you seem smarter, this would only help if people actually read it—which brings me to my next point.

The Selfishness of Writing Well

Achieving telepathic writing is hard work (chapter 2). I have spent many hundreds of hours crafting pieces of writing that I hoped might achieve crystal clarity, and in this book I will urge you to do the same. Those were hundreds of hours I could have spent doing more experiments, or drinking beer with friends, or even just walking along the water's edge skipping stones. So why invest the time and effort in writing well?

It might seem that working to make your writing clear is an act of selfless generosity toward the reader—this is the impression left by Strunk's metaphor of throwing the reader a rescue line. Or it might seem an act of generosity toward the progress of science. This was the argument made by Bacon, Sprat, and others in the 1600s; in this view, Newton was selfish in withholding his written work and writing for opacity. There's no question that writing well serves both the reader and the progress of science. But the evolution of science since Newton's time, and especially its spectacular growth, has changed the incentives for writing well.

In the 1680s, Newton had the luxury of writing a difficult book and knowing that every mathematician, physicist, and astronomer who mattered would invest whatever time was needed to grapple with his text. There just weren't many works of similar importance competing for their attention. But in our modern era, the deluge of published scientific work becomes greater every year. Just for the year 2012, for example, a Web of Science search returns more than sixty thousand records for oncology alone, fifty-one thousand for physical chemistry, and forty-five thousand for optics. By comparison, the mere nine thousand records for virology seem almost manageable—but even if you considered just ten percent of the virology literature relevant to your own work, keeping up with it would mean reading three papers every single day of the year. That *might* be possible for a while, but these numbers include only peer-reviewed papers in journals indexed by the WoS, not papers in more obscure journals, technical reports, books, book chapters, theses, grant proposals, or any of the other forms of scientific writing that form teetering piles in scientists' offices around the globe.

As a scientific writer, then, you are competing for attention with an incredible array of material your reader might prefer to your own. Your career and reputation, though, depend on having *your* work read. Hiring, promotion, and tenure committees and granting councils devour citation data for your publications. Grad-school admissions committees look for evidence of writing skill, and the best prospective graduate students search for supervisors by reading the literature to find someone whose ideas excite them. And, of course, journal editors and reviewers groan under the weight of submitted manuscripts, and can't be depended on to see the jewel hidden in a manuscript that's difficult to read. Readers have a lot to choose from. If your paper isn't clear they will turn to another. When they do, it's you as the writer who suffers most.

You can't make your reader like your science simply by writing better—but you *can* make it easier for them to see why they should like it, or at least why they should read and cite it. Given all this, the biggest winner when you put in the effort to make your writing clear is not your reader, and not the progress of science: it's you. And this is a victory you can shoot for, partly because there's so much bad writing out there for you to outshine (glass half-empty) and partly because you can practice your craft and learn to write better and better (glass half-full). Newton clung to a world in which the selfish act was to write opaquely, but in the modern world, scientists can do themselves no bigger favor than writing well.

The Transferability of Writing Skill

This book aims to help you improve your scientific writing. That you're reading it suggests that you plan such writing in the near future (if you aren't struggling with it now). But what if your career takes you away from academia and you never need to write a scientific paper again? Will the effort you put into improving your scientific writing be wasted?

In a word, no. Although I decorated my argument for the selfishness of writing well with details from the world of *scientific* writing, every bit of the argument holds for writing in other forms and other careers. Those who move away from scientific research may not write science after leaving the academy, but they will almost always write something else. Perhaps they'll complete a graduate degree in geology but then work in industry or government and write progress and technical reports. Perhaps after earning an undergraduate mathematics degree a student will go to law school and draft case summaries, legal opinions, or even legislation. Perhaps a biologist, fifteen years on, will end up writing instruction manuals, sales brochures, or—who knows—children's

TWO

||||||||||||||||||||||||||||||||||||||||||||||||||||||||||||||||||||||||||||||||||||||||||||||||||||||||||||||||||||||||||||||||||||||||||||||||||||||||||||||

Genius, Craft, and What This Book Is About

Genius vs. Craft

Writing comes naturally to some. Alexandre Dumas wrote *The Three Musketeers* and 276 more books, and in 1844 he famously made and won a bet that he could write the first volume of *Le Chevalier de Maison Rouge* in three days. Isaac Asimov wrote 506 books, and the romance novelist Barbara Cartland wrote 722[1]. George R. R. Martin (*Game of Thrones*) has written just thirty books so far, but they're really, really long. Writers such as these produce thousands of words of publication-ready text every day—a rate that seems possible only through genius.

Some scientific writers have this natural ability. Early in my career, I watched a senior scientist who seemed Dumas-like. To write a paper, he would walk around for a week or so with a thoughtfully tilted head, then sit down at the keyboard and let a nearly publication-ready draft drain out through his fingers into the computer. Like many new writers, I expected writing to be easy for me, too. I quickly learned otherwise. When I write, I stop and start; I write, delete, undelete, and delete again. I take a manuscript and reorganize, rewrite, rephrase, and polish it through a dozen drafts or more. Sometimes I spend hours on a passage, then throw it away and start over.

For a long time I thought my struggles made me unusual, but they don't. Most writers struggle. I didn't realize this because I had been seeing their writing *product*, not their writing *process*, which led to finished work that was clear, smooth, and easy to understand. Perhaps the iconic example of successful struggle is the nineteenth-century French novelist

---

[1] Depending on your feelings about romance as a genre, you may prefer to think of Cartland as having written one book 722 times.

Gustave Flaubert, who famously labored to find "le seul mot juste" (the only perfect word). He produced only a handful of novels in his career because composition was nothing short of anguish for him. (His best-known, *Madame Bovary*, took him five years to complete.) Flaubert wrote of once taking three days to make two corrections, and five days to write a single page—and yet he became revered as a great writer.

My realization that most writers, including scientific writers, worked hard at their craft, rather than being natural geniuses, was transformative for me. It led me to think of hard work at the craft of writing as a normal part of my job. In turn, this made it seem all right for me to spend the time—lots of it—it took to compose, revise, and polish. It also made me realize that I could set out deliberately to learn and to practice the elements of the craft, rather than sitting at my keyboard hoping for genius to strike. I saw that there were tremendous gains to be realized from conscious attention to my writing process and to my behavior as a writer, rather than just to the content I was producing. I found thinking this way tremendously empowering. Despite not being geniuses, you and I, and thousands of colleagues like us, can write very well—it just takes attention to the craft.

The Craft of Scientific Writing

But what does "attention to the craft" mean? Books about writing can seem mechanical and dull when they focus on the minutiae of grammar and usage:

> Use the present participle and present infinitive to indicate time that is the same as the time of the main verb, whatever the tense of the main verb is; use the perfect participle and the perfect infinitive to indicate time previous to the time of the main verb. (Johnson 1991:56)

> The mass number (of a nuclide) is shown as an anterior superscript: ^{14}N. A posterior superscript can indicate either a state of ionization: Ca^{2+}, or an excited state: $^{110}Ag^m$, $^{14}N^*$. A posterior subscript is used to indicate the number of atoms in a molecule: $^{14}N_2$. (American Institute of Physics Publications Board 1990)

This is not That Sort of Book. This isn't a book about synonyms, or grammar rules, or citation formatting, or table layouts. All those things are indeed part of the craft, and I'll address them where they're relevant, but this book isn't an exhaustive guide to them. Such technical matters are well covered in excellent (if often dry) guides that are widely available; I'll refer you to some as appropriate.

Instead, this book outlines a strategy for you, the ordinary scientific writer practicing your craft—a strategy with two elements. This can be applied through the entire process of scientific writing, from conceiving of a paper through revision and publication.

The first element is a relentless focus on the goal of crystal-clear communication: nearly every decision you make should be made with that in mind. Should you include a detail of methodology, or leave it out? Should you write in the active voice or the passive? How many decimal places should you give for the numbers in a table? Should your data be in a table at all, or in a figure? In each case, the route to an answer is the same: the better choice is the one that lets the reader more effortlessly understand the story you have to tell.

The second element is deliberate attention not just to *what* you write, but also to *how* you write. Many new scientific writers simply sit down and expect writing to happen (finding the process mysterious, as I did, if they think about it at all). Such writers can profit by consciously considering their own practices and behavior as they write. Engaging with yourself this way will let you write more, write more easily, and write better—although it does require honest discussion (even confrontation) with yourself about how you write.

To sum up: this book is about how writers can succeed by complementing attention to *what* they write with attention to their *goal* in writing (crystal-clear communication) and attention to the *way* they write (their process). Taking this deliberate approach is what makes writing a craft.

Describing scientific writing as a craft invites a comparison to another activity, such as cabinetmaking. In both, proficiency involves familiarity with basic materials (wood and fasteners, words and graphics) and their assembly into larger pieces with strength. Proficiency at either craft requires careful thought about the needs of users: who will use the product (be it a cabinet or a scientific paper), and how, and why? In both, skillful execution of technique is a result of dedicated practice. A cabinetmaker might practice drafting, cutting, joinery, and finishing. A writer might do exactly the same, only with words replacing wood as the material of choice. Finally, proficiency at either craft requires knowledge of one's own behavior, one's weaknesses and strengths, and how to manage one's efforts in order to produce more and better products. I can't teach you to be a cabinetmaker, but I think my advice can help you become a better writer.

Chapter Summary

- Few scientific writers are natural geniuses. Most find writing difficult and time-consuming.
- Scientific writing is a craft, and one can improve at the craft by practice and deliberate attention.
- Attention to the craft includes attention to writer behavior, not just written text.

Part II

||||||||||||||||||||||||

Behavior

"Writing" can be both a noun and a verb. As a noun[1], writing—and scientific writing in particular—is a form of expression with a history, a function, and a set of conventions. These are important, but their further treatment will have to wait for Part III. In Part II, we turn our attention to "writing" as a verb.

Thinking of "writing" as a verb draws our attention to writing as a process. I don't mean, of course, the mechanics of cursive writing or typing, but rather the intellectual activity of composition. A piece of writing doesn't appear out of nowhere. It is composed by a writer, and this composition is an activity that we can usefully think about. What are you doing, and what are you thinking, as you write—or, at least as importantly, as you *don't* write, even though you should? By talking about your behavior, I'm asking you to take your focus off *what* you're writing, and think instead about *how* you're writing it. Are you distracted? Do you write in short bouts or long ones, and how do you typically start and end a writing session? When you're stuck for a word, do you wait for it to come, skip it and move on, or take a break? Do you reward yourself for success in writing, and if so, when and how? These are all elements of writer behavior. Recognizing and modifying your writing behavior will help you improve your craft.

For most scientists, writing takes place behind closed doors. Nobody sees you write. You don't see anybody else write. It's the product, rather than the behavior, that gets shared. As a result, many scientists think of writing (if they think of it at all) as a mechanical process of transcribing thoughts onto paper. They may be aware of rules governing the writing-

---

[1] Strictly speaking, a gerund; but this sort of distinction is not what this book is about.

as-a-noun they are aiming for, but don't think much about the psychology and behavior involved in writing-as-a-verb. This is a very big mistake. Writing is produced by writers who write, and careful thought about you as a writer and *how* you write will be richly repaid.

What I have to say in Part II applies not just to scientific writing, but to any writing you do. However, you'll probably have scientific writing in mind as you read, as you'll see I largely did as I wrote. If you find that practice at scientific writing and practice at *non*scientific writing reinforce each other, so much the better.

THREE

||||||||||||||||||||||||||||||||||||||||||||||||||||||||||||||||||||||||||||||||||||||||||||||||||||||||||||||||||||||||||||||||||||||||||||||||||||||||||||||||

Reading

It's easy to think of writing and reading as two completely different activities, connected only by the passing of a printed page from writer to reader. Under this view, writers don't need to think about their readers: distribution channels such as publishers, libraries, booksellers, and the internet stand between the two and manage the necessary handoff. And, similarly, readers don't need to think about writing—which, by the time the reading happens, is safely in the past.

This view serves writers poorly. I have already argued that writers should have readers constantly in mind, because the goal for any writer should be crystal-clear communication with the intended audience. So, writers should *think about reading as they write*. But there is great value in making a connection in the other direction as well: writers should also *think about writing as they read*.

Reading with an Eye to Writing

The only way to produce crystal-clear writing is to know how a reader will respond to the choices you make in composing text and graphics. You need to know which words are most familiar, which sentence structures are most easily understood, which organization of material into sections is most easily followed, and so on. It's certainly possible to offer some general rules along these lines: for example, "use the active voice," "divide the paper into Introduction, Methods, Results, and Discussion sections," and "use a figure instead of a table when quantities are to be compared." In principle, you could tape a long list of such rules above your computer screen and treat it as the voice of authority on how to reach readers. But long lists of rules are boring, using them makes writing mechanical, and good writing sometimes entails knowing when to bend the rules instead of following them. Furthermore, using a list of rules is oddly indirect: instead of relying on rules you've been told will produce clear text, surely it would be more effective to understand how readers think, and write to that understanding?

Understanding how your readers think isn't easy, though (chapter 21). They will rarely tell you about their experience reading what you've written, and when they do, it's usually too late for you to change it. Fortunately, there's one reader you know very well, and who will talk if you're willing to listen. That reader, of course, is you. Your reactions to what others have written are exactly what your writing self needs to know about.

You can learn the most from yourself as a reader by paying deliberate attention to your reactions as you read. If you find a paper particularly easy or pleasurable to read, what made it so? What wording, structure, or graphics did you think were effective? If you found a paper hard, what elements made you struggle? Can you imagine a change that would have made the writing clearer? Steven Pinker (2014; his chapter 1) offers some concrete examples of this way of reading. Take written notes (or annotate the PDF) on examples of effective or ineffective writing and save them in a folder for later reference. When you write, think about your reading experience, imitate what you liked, and avoid recreating what you didn't. Actually, doing this deliberately is just an extension of what you've been doing subconsciously ever since you learned to read. Just as children develop an ear for spoken language by listening to their families, friends, and neighbors—and therefore speak with a vocabulary and accent that can pinpoint their origins decades later—so you develop an ear for written language by reading. Things you've liked as a reader will naturally crop up in your writing, but you can greatly accelerate the process with some conscious attention to the matter.

Once you've decided to pay attention to writing as you read, you can find opportunities everywhere. If you participate in a journal club, a lab meeting, or any other group of peers who read a paper together, make the writing of the paper, not just its content, an explicit focus of discussion. If you don't participate in such a group, start one. Offer to read manuscript drafts for your peers, or even better, for more senior colleagues. (Everyone needs "friendly review"; see Chapter 22.) If you're asked to provide peer review by a journal or granting agency, accept: you'll do a service to the profession, you'll build a relationship with an editor, and you'll have a chance to engage with someone else's (good and bad) writing. Finally, you needn't learn only from *scientific* writing. You can take hints from anything and everything: newspaper articles, blog posts, nutrition tables on your cereal box, trashy airport novels, or even *Wuthering Heights*. All of these use language to convey information and to persuade, and all can show you things to emulate and things to avoid. The more you read, of anything, the better.

But Be Careful about Plagiarism

While reading is a wonderful way to fill your writing toolbox and a great source of writing inspiration, be careful not to stray over the line into plagiarism. Plagiarism is the presentation, whether deliberate or accidental, of someone else's words, data, graphics, or ideas as if they were your own. My recommendation that you read and then imitate what you admire does not, of course, extend to the appropriation of content from your sources. (There is a cultural context to plagiarism and its definition that can trip up scientific writers from non-Western cultures. More about this in chapter 27.)

Most scientific writers are not in the business of deliberate plagiarism[1], so what I'm cautioning against here is inadvertent transgression. You can avoid crossing the line with some careful attention to three issues, one or more of which lie at the root of most cases of inadvertent plagiarism. First, content differs from style: you are free to imitate the way another writer said something, but not what they actually said. This means that if you admire a writer's organization, style, turn of phrase, or graphical formatting, you may generally imitate it without worry; however, you may *not* copy (without attribution) wording above a short

---

[1] At the undergraduate level this unfortunately may not be true. It seems likely that a majority of undergraduate students have copied at least some academic work (Blum 2009). Even at the graduate and professional levels, deliberate plagiarism is far from unknown, and a quick search on the internet will turn up the latest case in your discipline. However, I will assume that writers serious enough about their craft to be reading this book are not the ones tempted to short-circuit the whole process with theft by mouse click.

phrase, data, graphical elements such as maps or drawings, or other substantive pieces of text or graphics. Second, remember that paraphrasing the words of another writer means more than changing a few words, or even changing all of them while leaving ordering and phrasing intact. A good paraphrase uses your own independently designed wording and phrasing to express the content you admired from elsewhere (but that must still be attributed via citation). Third, when you save examples of good writing so you can imitate them later, be meticulous about identifying their sources, even in your own informal notes—lest you later mistake something you wish you'd written for something you actually did, and incorporate it verbatim into your work. Perhaps this last piece of advice sounds too obvious to be worth giving, but a Google search for "plagiarism sloppy notes" returns (as I write this) nearly nine million hits!

Further advice on plagiarism is widely available. Pecorari (2008) considers plagiarism as a linguistic phenomenon and discusses intentional vs. unintentional plagiarism. Comparisons of acceptable paraphrasing and plagiarism are provided by writing centers at many universities (e.g., http://www.indiana.edu/~wts/pamphlets/plagiarism .shtml). A more technical guide to plagiarism in academic publication, "Authorial Integrity in Scientific Publication," is available from the Society for Industrial and Applied Mathematics (http://www.siam .org/journals/plagiarism.php).

Reprise: The Value of Reading

Don't let cautions about plagiarism deter you from taking advantage of your reading. The ear you develop from reading (especially with conscious attention to writing) is worth more than a thousand writing rules. As a writer, you are part of a community stretching back thousands of years, and what's been written before you can be thought of as a long series of experiments in writer-reader communication. Reading, whether of science, literature, or cereal boxes, gives you access to the results of those experiments and lets you apply them to your own writing craft. So read: read often, read broadly, and read deliberately.

Chapter Summary

- Reading is a useful way to build writing skills.
- While reading, make notes about writing you find successful or unsuccessful, to model or avoid yourself.
- When modeling your own work after good writing, be careful to avoid plagiarism.

Exercises

1. For the next three scientific papers you read, make written notes of at least one aspect of the writing you admire, and at least one aspect you think could be improved.

FOUR

|||||||||||||||||||||||||||||||||||||||||||||||||||||||||||||||||||||||||||||||||||||||||||||||||||||||||||||||||||||||||||||||||||||||||||

Managing Your Writing Behavior

I've always been a slow writer. It's not uncommon for me to devote an entire day to a writing project and end up with just two or three paragraphs of new text. The first draft of a manuscript can take me weeks, if not months, of work. By itself, this isn't a terribly important thing: many successful writers have been notably, even famously, slow[1]. What's more important is that there are two reasons I might be slow on any given day. Some days I'm slow because I spend the whole day writing and rewriting, deleting and undeleting, phrasing and rephrasing, organizing and reorganizing. You may think of this as "Flaubert slow" (chapter 2). Other days, though, I'm "dog-sees-squirrel slow": repeatedly distracted, and my own worst enemy. For instance, after opening the blank document that was to become this chapter, but before writing anything past the title, I checked my e-mail four times, read news articles in the *New York Times* and the Toronto *Globe and Mail*, went to the greenhouse to weed (unnecessarily) goldenrods growing for an experiment, read the latest postings on a baseball blog, a computer-security blog, and two economics blogs, and thought hard about whether it was close enough to noon to heat up my lunch. (Sadly, it wasn't.) None of these things helped add words to the page.

My two ways of being slow arise from very different behaviors. They have different effects on the quality and quantity of writing I produce,

---

[1] None perhaps as resolutely slow as Shelby Foote, who took twenty years to write his three-volume history of the U.S. Civil War (Foote 1958, 1963, 1974). Foote wrote longhand with a dip pen and inkwell (well into the 1970s!), blotting wet ink from every page, and didn't write down anything he didn't like because he hated having to revise. Unsurprisingly, Foote considered it a good day if he produced five hundred words (Foote 1994).

and to the extent that they are problems, they have different solutions. Through my career as a writer, I've become more productive partly because I learned to recognize the behaviors underlying my slow writing and to find ways to change them: to revise when revision was useful and avoid it when it was wasted effort, and to recognize when I was chasing squirrels and get my attention back on the blank page. This is just my own example of a more general principle: as you practice your craft, it's essential that you pay explicit attention not just to the content of what you are writing, but also to your own behavior as you write it.

Why Behavioral Self-Awareness Matters

It is surprising how many writers overlook the obvious: you can't write more or write better without changing what you're doing as you write, and you can't *change* what you're doing unless you *know* what you're doing. In this respect writing behavior is no different from nail-biting, overeating, or any other behavior one might like to modify. The first step in managing one's behavior is making a deliberate decision to be conscious of that behavior. Trivial as this step sounds, it took years for me to take it, and the world abounds with writers who haven't done so yet.

The benefits of self-awareness, though, go far beyond letting you decide to change your behavior. As it turns out, it can also help you *implement* and *maintain* such a decision. We know this from neurobiological studies using a technique called functional magnetic resonance imaging (fMRI) to detect conscious self-reflection (Box 4.1). In such studies, participants are asked to improve their performance at a task (something like stopping smoking; but the logic applies to writing too). Those who are most successful, and successful longest, are the ones who think consciously about their own behavior and how it relates to the task.

Self-awareness doesn't just help while one remains self-aware, either: it carries on helping sustain behavioral changes much later. This is a good thing, because maintaining awareness of your own behavior is not easy. Most of us are very good at slipping into unconscious execution of a familiar task. (If you doubt this, go for a walk and see how long you can pay explicit attention to what your toes are doing.) Not only that, but the very attention paid to a behavior can interfere with its execution: while thinking about your toes, you may walk into a telephone pole. If the only way to manage your writing behavior was to be consciously aware of it *all the time*, this wouldn't be a very fruitful avenue for improvement. Fortunately, the fMRI work confirms that this isn't so.

Box 4.1 Functional MRI and the benefits of self-awareness

Functional magnetic resonance imaging (fMRI) is a technique that can identify areas of the brain with elevated neural activity during a given mental task. It does this by using a contrast in magnetic properties between oxygenated and deoxygenated blood to measure energy consumption by brain cells. fMRI scans indicate that a few small areas of the brain are heavily involved in conscious thinking about the self. Activity in the dorsomedial prefrontal cortex (dmPFC), in particular, is associated with evaluation of, and decision-making about, one's own behavior (van der Meer et al. 2010). But dmPFC activity isn't just a convenient indicator that someone is thinking about their[2] own behavior: it also predicts the person's ability to change that behavior in the future. Chua et al. (2011) used fMRI to monitor brain activity in people who wanted to stop smoking, while they listened to different types of smoking-cessation messages. As expected, the messages that caused the most dmPFC activity were those that were tailored to the test subject's individual background and circumstances (health history, self-identified reasons for smoking, and so on)—messages that provoked them to think consciously about their own behavior. Smokers who had heard these dmPFC-activating messages had more success quitting than others who heard more generic messages about the desirability of quitting. Chua et al. (2011) speculated that this effect arose because self-reflective thinking stimulates deeper neural processing related to the smoker's goal of behavioral change, and that this leads to better integration of the goal into learned (changed) behavior. Importantly, dmPFC-activating messages were associated with changes in behavior long after those messages were actually heard—that is, weeks or months after the subjects did their self-reflective thinking. Thinking explicitly about their own behavior at one time (while listening to the

---

[2] Throughout this book, I avoid the noninclusive "he" and "she" and the awkward "(s)he," "he or she," and other such attempts to overcome the defect in a language that lacks a nongendered third-person singular pronoun. Frequently, I adopt the singular "they." While this is often held to be a grammatical error, there is a good case for it, both historically and functionally. Those interested can refer to my blog post on the subject: http://wp.me/p5x2kS-39.

messages) helped smokers make and sustain behavioral changes into the future. This understanding of how your brain works applies much more broadly than to smoking, and I think you can exploit it to better manage your behavior as a writer.

An interesting corollary to the importance of self-awareness is that general advice offered in a writing book—even if it happens to be perfectly suited to you as an individual writer—will be less effective than advice you give yourself as a result of thinking about your own behavior. This doesn't make this book unimportant, but it does underscore the fact that you must be an active, thoughtful partner in my effort to help your writing craft. If sometimes I fail to tell you what to do, and instead only tell you that *you* need to decide what to do, it isn't a cop-out: not only are you unique as a writer, you have a lot to gain from thinking about your own uniqueness.

Encouraging Behavioral Self-Awareness

Behavioral self-awareness is easier suggested than achieved. It's possible to use some simple tricks to help bring your own behavior periodically to mind. One of these may work for you:

- **Reminders**. Put a little sign above your writing station (laptop screen, desk, etc.) that says "*How* are you writing?" When your gaze crosses the sign, you'll be asking yourself to notice your behavior. Are you actually writing, or are you distracted? Are you writing new text or revising old? Whichever it is, is that what you should be doing? Try to prevent the sign from fading into your cognitive background: move it around from day to day, print out a new one with a different font, or do something else to keep it fresh.

  Alternatively, hang a small stuffed animal or the like near your writing station, and think of it as your writing conscience: imagine that it's observing your behavior as you write. When you notice it, think about what it's seeing, and you'll see the behavior too. I use a little Pinocchio puppet (which has the additional advantage of reminding me that what I write should mostly be true).

- **Writing logs**. Reminders work in the moment; another approach is to force yourself to think retrospectively about your writing behavior. Writing logs can help focus this thinking. One approach is to document an individual writing session. Choose a writing session of two to three hours' length and keep a six-minute writing log: that is, set an alarm to go off every six minutes, and when it sounds, jot down what you were doing at that instant. You may be surprised at the fraction of entries that involve staring off into space, checking e-mail, fetching a snack, or something else other than writing (Figure 4.1). If so, you've identified some behavior to manage.

  The six-minute log isn't something you'll want to implement often, though, because it risks interrupting your writing and causing the very problems it's supposed to diagnose. (There are online tools that compile logs in the background, such as RescueTime [http://www. rescuetime.com], although these can only track your use of internet-connected devices.) A good complement that you can keep up in the longer term is a daily writing log. At the end of your writing day, spend five or ten minutes writing a few notes about the day, and your behavior as you wrote. How much did you write, and did you write something of high quality? If so, what did you do that made that possible? If you didn't accomplish much, what got in the way, and how could you avoid that problem the next day? Don't just think about this; actually writing it down will help focus your thoughts and give you something to revisit in the longer term, to look for patterns and draw lessons.

- **Cooperate with a friend**. Agree to discuss (regularly) writing behavior with a friend or colleague. For instance, you might agree to send each other copies every evening of what you wrote or revised that day, along with some commentary on what you actually did while you were writing. Alternatively, you could simply agree to exchange and discuss the daily writing logs suggested in the previous bullet. If your friendship can stand it, go one better: agree to check each other's browser histories, Facebook pages, refrigerators, and other evidence of behavior during writing (or non-writing!). This strategy has two advantages. First, the involvement of another person can act as a commitment device (chapter 6), making it harder for you to skip the day's self-reflection. Second, another person can often identify a be-

At:	I was:	At:	I was:
9:00	writing new text	10:00	writing new text
9:06	writing new text	10:06	writing new text
9:12	writing new text	10:12	writing new text
9:18	checking e-mail	10:18	fixing snack
9:24	writing new text	10:24	reading legal humor blog
9:30	fiddling with wording	10:30	writing new text
9:36	reading baseball blog	10:36	checking Twitter
9:42	checking e-mail	10:42	writing new text
9:48	writing new text	10:48	staring into space
9:54	reading *Dilbert*	10:54	writing new text

Figure 4.1 A 6-minute writing log for one of my writing sessions. Notice that I was actually writing for less than half the time!

havior you aren't even aware of, or ask you a question you wouldn't have thought to ask yourself.

Of course, none of these suggestions will work without some painful honesty. It can be quite a shock to compare what you think of yourself as doing (writing) with what you actually do (updating Facebook, making paper-clip chains). You can at least take heart that any bad habit you discover is sure to be one that another writer before you has shared and overcome.

Some Common Behavioral Challenges

I argued earlier that every writer is unique, and that there is value in discovering your own behavior for yourself. Nonetheless, any list of behavioral challenges for writers will probably start with the following Big Six. If you recognize one of these in your own writing behavior, you are in excellent company:

- **Avoidance.** Many writers will put off starting a writing project, or sitting down to a day's writing session, for as long as they can. The blank page can be intimidating, but the only way to get rid of a blank page is to start filling it. There are, however, ways to make starting a project or a session easier (chapters 5 and 6).

- **Distraction**. The opening paragraph of this chapter should identify pretty clearly my own behavioral challenge: I have no difficulty sitting down to write, but an enormous problem staying there. Without the ability to maintain focus on the activity of writing, it's really not possible to produce the volume of written text needed for a successful career. I've learned a lot about distraction in my career as a writer (chapter 6).

- **Feeling stuck**. Perhaps your writing pauses not while you're distracted by something else, but while you stare at the page without knowing what to write next. This problem is often referred to as "writer's block," but that's a poor name for it, because it suggests a force that comes from outside the writer, instead of a behavior that comes from inside. Overcoming "stuckness" is a matter of maintaining writing momentum (chapter 6).

- **Perfectionism**. It seems as though it should always be a good thing to want your writing to be better, but as in most things, a little moderation is wise. A perfectionist who won't turn to the next paragraph until the preceding one is perfect is unlikely ever to reach the last paragraph. There is much to be said for getting something lousy on paper (chapter 6) and fixing it later (chapter 21).

- **Fear of criticism**. Early-career writers are sometimes reluctant to show their work to colleagues or (even more) to supervisors, because criticism stings. Instead, they polish and polish, but never decide the work is good enough to share. Criticism does sting, to be sure[3]. Successful writers realize that the sting is useful: the whole point of writing is to communicate with a reader, and the only way to judge and

---

[3] Every writer gets criticism. As a student, novelist and Oscar-winning screenplay writer William Goldman coedited Oberlin College's literary magazine. He submitted his own stories anonymously and remembers his coeditors saying of them, "We can't possibly publish this shit" (Queenan 2009). Fortunately for fans of *The Princess Bride* and other classics, he stuck with the craft. Margaret Mitchell's manuscript for *Gone With the Wind* was rejected thirty-eight times, but the published version won a Pulitzer; Madeleine L'Engle's *A Wrinkle in Time* was rejected twenty-six times before the published version won the Newbery Medal. None of these writers enjoyed the criticism, but the work of each was improved by it. My own most-rejected paper (Heard and Remer 2008) was turned down by six different journals. The published version of that paper, as a direct result, is completely different and much, much better than the original draft.

improve that communication is to share draft writing and accept criticism (chapters 22 and 23).

- **Reluctance to revise**. Criticism of your work is invaluable, but only if you put it to use. A common tendency is for writers to resist making suggested changes to their writing. I see this all the time in my role as an editor, and I struggle with it as a writer. There are some natural psychological forces behind this resistance, and so most writers need to pay deliberate attention to overcoming it (chapters 22 through 24).

Your own behavioral challenges may not be on this list, but you're sure to face something. If you would like to write more than you do, or better than you do, think not just about the content of what you are writing, but also about yourself as a writer. Understanding and managing your own behavior is the key to actually accomplishing all those things that you know you are supposed to—all the way from correct punctuation to career-long productivity.

Chapter Summary

- Understanding and managing your own writing behavior is essential to productive writing. Each writer will have a unique set of behavioral challenges.
- People who are consciously aware of their own behavior are better able to manage it. Tools for maintaining behavioral awareness include posted reminders, writing logs, and monitoring agreements with friends or colleagues.

Exercises

1. Plan a two-hour writing session and make a six-minute writing log. What writing behaviors did it identify that might be reducing your productivity? What other unproductive behaviors do you suspect of yourself?

FIVE

Getting Started

Because I struggle at writing, I don't enjoy it very much. As a result, when it's time to begin a new piece of writing, I find it hard to get started. Almost everything else on my to-do list seems preferable, and a lot of things not on the list seem pretty alluring as well.

Of course, reluctance to write can strike at any time, but many writers find starting a new writing project much more daunting than continuing once they've begun. John Steinbeck, for instance, confessed in his journal, "I suffer . . . from the fear of putting down the first line. It is amazing the terrors, the magics, the prayers, the straightening shyness that assails one"—and if he had trouble starting, it can't be surprising if you or I do (Steinbeck 1969; 13 Feb. 1951, 9). With the whole project laid out before you, the hardest parts draw your attention like the highest hills on the horizon, and it's hard to picture ever reaching the end. So the first day's work ends up slipping, perhaps just a day or two, or perhaps a little more. The problem, of course, is that most scientific writers need to start a lot of different writing projects—papers, chapters, proposals, reviews, reports, and so on. A writer who loses just a few days to difficulty starting each new project could fall months behind in a single year. So, if you have trouble getting started, the issue is worth attention.

With chapter 4 fresh in your mind, you will not be surprised to find us beginning with some conscious attention to writer behavior. There are two major ways in which writers fail to get started, which I'll call "unintentional" and "intentional" non-starting. Begin by identifying the source of your own hesitation. Intentional non-starters make deliberate decisions not to start writing (yet). Unintentional non-starters (like me) make no such decision but nonetheless reach the end of their working day and discover they haven't started. These two types of writers end up in the same place—not having started—but their distinct behaviors need separate treatment.

Unintentional Non-Starting

Unintentional non-starting is really just a form of procrastination. To procrastinate is to "delay an intended course of action despite expecting to be worse off for the delay" (Steel 2007, 66). This has been recognized as a widespread human failing for thousands of years[1]. Actually, procrastination isn't just a human behavior: Mazur (1996) demonstrated that pigeons do it too.

We all know we shouldn't procrastinate, and yet we all do. So what keeps us doing it, and can we change our behavior? The psychological literature brings both bad and good news. The bad news is that some people are genuinely more vulnerable to procrastination than others (Steel 2007), with their vulnerability persistent through time and perhaps even under genetic control (Arvey et al. 2006). The good news, though, is that this is a tendency, not inescapable doom: people can and do learn to reduce procrastination. Even better, understanding the psychological roots of the behavior can help you find ways to control it (Steel 2007).

It's useful to think of procrastination as the outcome of a choice you make between two or more behaviors (or tasks) that you might choose to begin at a given moment. Imagine, for instance, that right now you could start to write, check your e-mail, or get up to fetch a snack. Each task would deliver rewards during execution or at completion, but these rewards differ in value and are expected at different times in the future. The key insight that emerges from psychology (Steel 2007) and behavioral economics (Thaler 1981) is that humans don't just choose the task with the largest reward. Instead, we are less motivated by rewards expected further in the future: technically, we "discount" their value. Thus,

---

[1] The Roman orator Cicero railed against procrastination in 44 BC (in his sixth *Philippic*), as part of a scathing attack on Mark Antony. The earliest known Western reference is from the ancient Greek poet Hesiod, who counselled against procrastination in his tiresomely moralistic 828-verse poem *Works and Days*. Farther east, in the *Bhagavad Ghita* Krishna lists procrastination along with vulgarity, laziness, wickedness, and deceitfulness as sins defining "Taamasika" persons who are fated, upon death, to be reborn only as beasts or insects. Ouch.

we tend naturally to choose a task offering a small immediate reward (say, reading that newly arrived e-mail) over a task with a much larger deferred reward (say, the satisfaction of finishing a manuscript).

This perspective is formalized in temporal motivation theory (TMT; Steel and König 2006), and the formalization is helpful because it can give us a coherent framework to unite and understand behavioral strategies for avoiding procrastination. TMT models task choice as the outcome of a calculation (explicit or implicit) of the attractiveness of each task, with the most attractive being chosen. Task attractiveness depends on four major factors: how confident one is of completing the task ("expectancy"), the importance one ascribes to the reward ("value"), how much one devalues future rewards ("discounting"), and how long it will be before the reward is received ("delay"). Each of these factors depends on properties of the task, but also of the person choosing. More specifically,

$$A \propto \frac{E \cdot V}{\Gamma \cdot D}$$

where A denotes attractiveness, E expectancy, V value, Γ discounting, and D delay.

Let's work through this expression. While there is controversy over the shape of the functions involved (especially the form of discounting implicit in Γ; Rubinstein 2003), the qualitative influence of each term is fairly intuitive. From the numerator, we see that attractiveness *increases* with expectancy: if you are more confident of completing the task, you will be more likely to start. Attractiveness also *increases* with the expected value of the reward (say, its importance to your career). From the denominator, we see that task attractiveness *decreases* with discounting. Γ measures how strongly one's interest in a reward pales as the reward slips into the future: you have a high Γ if you tend to be impulsive, and a low one if you tend to weigh future consequences carefully before you act. Task attractiveness also *decreases* with delay, as rewards further in the future are devalued more by your application of discounting.

Importantly, E, V, Γ, and D aren't entirely beyond your control. You can manipulate each of them, with respect either to individual writing tasks or your writing behavior more generally. Taking action to increase expectancy or value, or to decrease discounting or delay, increases the attractiveness of writing and thus the likelihood that you'll choose that behavior. Taking them in turn:

Expectancy (E). There are at least three ways to increase expectancy for writing.

First, you can make writing seem easier. The intrinsic difficulty of a particular writing task is out of your control, but you can still manipulate expectancy by redefining the task. Do you (quite accurately) perceive writing the finished version of a paper as very difficult? Then focus instead on producing a rough draft, which is a much easier task. Is an entire draft still daunting? Then focus instead on the first section. This redefinition increases expectancy because the smaller task can be completed more easily (and sooner, which as a bonus decreases delay). Alternatively, invite a coauthor to join your writing project (see chapter 26). With a coauthor, you can divide the writing, each taking on the tasks you're most confident about (highest expectancy).

Second, you can work to get better at writing. Of course, you can't become a master craftsperson overnight, but every improvement you make increases expectancy and thus the attractiveness of starting to write.

Third, you can think more highly of yourself as a writer, even if your actual ability remains constant. Most obviously, you can remind yourself that you're capable of success. If you once wrote something you're proud of, reread it, and know that your new project can reach the same heights. Observing the writing success of your peers works, too (Bandura 1997): if they can do it, you can too. Even better is to reverse the logic, seeing successful writers as your peers. If you learn that scientific writers you respect, or famous writers like Flaubert (chapter 2), struggled just like you but still succeeded, your confidence can only be boosted—and with it, the attractiveness of diving into the job.

Value (V). Can you increase the rewards associated with writing? You can't easily manipulate rewards others give you (grades, career recognition, etc.), but you can bribe yourself a little. Might completing your manuscript earn a day off? Could each thousand words of a draft be worth a piece of good chocolate? Such things will combine with external rewards to increase total value. Of course, self-rewards should only come if you're actually performing the task; if you're tempted to cheat, consider asking a friend to dispense them. Training by sufficiently frequent rewards can even make performing the task pleasurable in its own right, via classical conditioning (Eisenberger 1992).

Another approach is to manipulate task aversiveness (the unpleasantness you associate with working at a task). Task aversiveness can be incorporated into the TMT model as a "reward" of negative value. Reducing the unpleasantness of writing thus increases value and makes procrastination less likely. You might want to spend a little extra money on a comfortable office chair[2], keep a scented candle at your desk, or stock the refrigerator with your favorite soft drinks.

Discounting (Γ). The discounting term in the TMT model largely captures attributes of the person choosing tasks rather than of the tasks themselves. Let's get two bits of bad news out of the way. First, research in psychology shows that variation among individuals in impulsiveness (high Γ) vs. mindfulness of future rewards (low Γ) tends to be fairly stable (Gustavsson et al. 1997). Second, the extensive self-help literature on impulsivity mostly boils down to earnest but unhelpful advice to stop being impulsive. Fortunately, all is not lost, because TMT theory suggests (and data confirm) that people discount less when they're consciously aware of future rewards. That is, conscious choices tend to set a smaller value for Γ (favoring high attractiveness of future rewards) than do unconscious ones. You can increase the attractiveness of writing, then, by encouraging awareness of the rewards to come. If you have a job offer contingent on defending your thesis, display it over your computer screen. If you are writing a journal paper, display a mockup of how its title will appear on the cover. Ask yourself why writing matters to you (what rewards you expect) and then be conscious of that.

Delay (D). If procrastination happens because we discount future rewards, can't we move the rewards closer (reduce delay)? Many re-

---

[2] Voltaire, the eighteenth-century French satirist, apparently liked to use his lover's naked back as a desk (Ackerman 1990). No word on what he used for a chair.

wards for writing can't be manipulated this way: they don't accrue until the work is finished, and/or they're delivered by entities like grant panels that keep their own schedules. However, you can control the timing of self-rewards. Because rewards that come early in the writing process (small delay) have outsized influence on your decision to start or to procrastinate, it's generally more effective to offer yourself small rewards for incremental progress than to promise yourself a trip to Disneyland when the task is completely done.

Long and uncertain delays outside your control may be especially damaging to your motivation. For example, I find that the months between submitting a manuscript and (I hope) having it accepted for publication seem to make it hard to focus on the rewards of writing. My fix: I promise myself that when I hit the "submit" button on a journal's website, I'll take the rest of the day off. Although this reward isn't great, it has a small and known delay, and I find it highly motivating.

Of course, if you can manipulate E, V, Γ, and D to make writing more attractive, you can manipulate them to make other tasks *less* attractive, too. Perhaps the best example is reducing electronic connectedness. Rewards for checking e-mail or Facebook, for example, come almost instantaneously, making those tasks highly attractive. Something as simple as writing on a computer with WiFi turned off can have an astonishing effect on your productivity, by forcing delay and discounting of such electronic distractions' rewards. Allowing yourself access only to video games that you dislike (low value) or that you're not good at (low expectancy) is a similar manipulation.

Finally, by representing procrastination as a choice, the TMT model suggests another approach. Rather than trying to ensure you'll make the "right" choice, you can see choice itself as the problem. You can remove the choice, and so make sure you start writing, by removing from your environment any cues indicating that distractions are available to you. Don't just leave your e-mail program closed (increasing delay to decrease attractiveness); delete its icon from your desktop (avoiding the temptation to calculate attractiveness at all). Free your desk of half-read thrillers, smartphones, and other temptations. Alternatively, you can avoid choices through automaticity: make writing a habitual activity that you maintain without conscious attention. If you can become used to the idea that in *this* chair, or at *this* time of day, you write and only write (Silvia 2007), then you've removed the choices that lead to procrastination. If it's that time and you need to start a new writing project, then start you will.

Intentional Non-Starting

While unintentional non-starters—procrastinators—know that they *ought* to start writing, intentional non-starters are a different breed. Intentional non-starters are sure that they will write faster and better if they wait for the right moment. Nearly always, though, they are wrong.

If you ask intentional non-starters what they're waiting for, you'll hear two very popular answers:

"I don't have all the data/analyses yet." You definitely shouldn't complete your Results section before you've gathered any data, but waiting to start until every bit of data is in hand and analyzed isn't a good idea either. You can begin writing at least as soon as you have enough data to know what overall story you plan to tell (and even sooner is better; see "Early writing," below).

Some intentional non-starters worry that if they begin writing too soon, the last analysis will change the story, and their effort will be wasted. This worry is misplaced. If analyzing that last piece of data drives you to rewrite two paragraphs of the Introduction, that task will hardly be noticeable among all the other revision you'll be doing (chapter 21)—but it will happen sooner, because you got an early start on the work. Even the notion that you can identify the "last" piece of data or analysis before writing is misguided. People who read your manuscript during friendly and formal review (chapters 22–23) routinely suggest the inclusion of new data or analyses, often to the dramatic benefit of the manuscript. Your own reading of a draft, or the thought about your work that's required to write it, can suggest an analysis or a piece of missing data. If you take advantage of it, there can be a powerful feedback between the developing manuscript and the science it will report: not only does the science suggest the story to tell, the developing story suggests directions for the science to go. Waiting to write until all the data are in hand and all analyses are done means missing this opportunity.

"I don't know what I'm going to say yet." Many intentional non-starters have a peculiarly static view of writing as a mechanical process, in which you simply manipulate a pencil or keyboard to record in detail the complete story already worked out in your head. Writers like this picture themselves drafting their papers mentally, and intend to wait to put pen to paper until they can simply transcribe that complete mental draft. If this were possible, then starting too early would indeed waste effort by forcing revisions later. But almost nobody can actually write like that. Nearly all of us use a much more dynamic process, in which we explore our story during the act of writing and our draft takes twists and turns as we revise broadly, deeply, and repeatedly (chapter 21).

Don't get me wrong: every writer lets at least some thought precede the actual writing. You might try out structures and arguments in your head, or rehearse passages to yourself. There's nothing wrong with that! But keeping it up for too long is self-handicapping. If you like some text that you've composed in your head, you need to do two things: remember what you've composed, and see if it works as well for a reader on the page. Both needs are best satisfied by writing down your thoughts. But then, if you're going to write down "finished" passages, why not incomplete ones? Why not scribble outlines, or jot down thoughts, or try out sentences or paragraphs as you type? That is, why not just write?

Other writers succumb to an even more dangerous version of prewriting composition. These writers may have gathered and analyzed all their data, but they aren't writing, and they aren't actively thinking about writing. Instead, they're waiting for some mysterious process to "inspire" them with a manuscript ready for transcription onto paper. The magic may involve the brilliance of their subconscious or gifts from writing pixies, but it's not something they can directly control. Writers dependent on magic always find themselves, sooner or later, abandoned by it. At this point, they are forced to struggle like the rest of us—except that they've lost time waiting for the magic that didn't come.

If magical writing inspiration does happen to you, of course, you should take advantage of it[3], but don't be tempted to depend on it. Successful writers write what they need to write, when they need to write it. In doing so, they discover that writing when you don't feel ready is a valuable aid to rigorous thinking. Putting words on the page does something for you that won't happen inside your head, no matter how long you wait for it.

"Early Writing": Integrating Writing with Doing Science

What if you took my advice from the previous section to its logical extreme? What if you never "started" writing up a project because you were writing all the way through its planning and execution? We can call writing that's integrated with planning and conducting of research "early writing." Adopting the habit makes doing science and writing about it threads of a single interwoven activity. It's a way to ease the writing process while also tapping into opportunities for synergy between writing and doing.

The advantages of early writing are most obvious with respect to a Methods section. First, there's no easier time to write up your methods than when you're planning or executing the work (and haven't yet forgotten any details). Second, writing the Methods can strengthen your experimental design by alerting you to ill-advised features before it's too late. Writing out your methods for an unfamiliar reader will shine a spotlight on gaps in your logic, mismatches between hypotheses and data, or missing observations that would make your story complete. In particular, watch for features of your methods that are difficult to explain. If you find yourself writing a convoluted explanation, or defending something more than explaining it, these are strong hints that you haven't chosen quite the right procedure or analysis.

Other sections are suitable for early writing, too. A natural time to draft material for the Introduction and Discussion (or to remodel mate-

---

[3] I have a SCUBA notepad in my shower, because I frequently get ideas of magical origin there. Shower magic has brought me ideas for experiments, elegant turns of phrase, and better ways to organize passages—but only haphazardly. I can turn the shower on and off, but the magic comes only when it pleases.

rial from a grant or project proposal) is while you're reading the literature and thinking about the field in order to plan your study. If you write
brief summaries of papers as you read them, for instance, you can test
your understanding and also end up with draft material for the literature-
review component of your own paper. And your project presumably
arose because you identified a knowledge gap in your field, so writing
down your rationale for the work will give you a draft of a key component of your Introduction (chapter 10). If this rationale doesn't stand up
to the rigorous thought forced on you by writing, it's better to find out
before you've spent time and money on work you can't sell to readers.

Even large parts of the Results section can be drafted before you have
actual results to put in it. It's very helpful to make mockups of the tables
and figures in which you imagine reporting your data—using pilot data
if you have it, and what we might call "simulated" data (invented data of
the sort you might expect from your work) if you don't. You can do the
same for your planned statistical analyses. Doing this as early as possible, ideally before you've taken a single measurement, is an excellent
way to test-drive the design of your study. More than once I've discovered through mockups a feature of my design that would have made
data analysis or presentation needlessly complicated—fortunately, in
time to fix the problem. Perhaps you are uneasy with the notion of making up "data," but remember, the simulated data serve *only* for this test-
drive and will never see the light of day. (Always label each mockup with
a "**simulated data only**" banner to avoid confusion later.) When you've
gathered the real data, you can cut and paste into the mockups and be
well on your way to a final version.

Of course, much of what you write before or during the work itself
will need to be revised, rewritten, or even (gasp!) discarded later on. It's
common, if not routine, to rethink your work as your data bring you
surprises. So is early writing wasted effort? Not at all. Revising drafts is
always much easier than writing from scratch, and even early writing
that's completely discarded will have paid dividends in helping polish
the science itself. Of course, it *would* be wasted effort to polish your
early writing, so feel free to write ungrammatically or in point form, to
ignore ugly formatting in mockup figures and tables, or to write discon-
nected blocks of text that you can fit together later. Shortcuts like these

reduce the cost of early writing, while retaining the enormous benefits of integrating writing with doing science.

Easing In

Getting into a cold swimming pool is easiest at the deep end, but getting into a writing project is easiest at the shallow end. If starting seems difficult, you can fool yourself by easing into the job. When you've decided to "start" assembling a new manuscript (even if early writing means you already have a fair bit of material drafted), begin with whatever part of the project you find easiest. Don't worry if this means what you write first isn't the part that will be read first: by the time readers see your work, they won't know or care where you started. When I write a paper, for example, I start with the acknowledgements. Sure, these are trivial, but once they're done, I feel as though I'm underway. Next I take a slightly harder step, perhaps drafting a figure or a table. Before long I'm sweating over the right sentence structure in a hard section. Easing in doesn't change the fact that writing is hard work, but using an easy bit as a gateway at least gives you momentum that sees you doing the harder stuff.

Not everyone finds the same bits easy, of course. An early writer might "start" by assembling and connecting pieces of Methods text written before and during execution of the work. Others find tables and figures a seductive place to start, because they emerge naturally from statistical analysis. Still others enjoy the literature review part of the Introduction, especially because it can cannibalize text from an already-written grant or project proposal or from those summary notes you made as you read papers for background. Another common strategy is to ease in by making an outline or concept map (chapter 7). Finally, one colleague of mine starts with the first paragraph of the Introduction (something that would be utterly toxic to me), but she has a trick: she takes the first paragraph of a paper she thinks is well-written, and then phrase for phrase and sentence for sentence, replaces the text with her own story. It doesn't matter if the paper she's emulating is on a completely different topic; what she's after is the clarity of writing and organization. This is much less frightening than facing the blank page, and even though she always rewrites the draft later, it eases her into independent writing once she's run out of text to model.

You can ease into any writing task, not just a new paper you're starting from scratch: a section, a figure, a paragraph. You can even ease into a sentence; I often bash out just a few typo-ridden words before reshaping them into a coherent sentence, and that way the eventual need for coherence doesn't scare me off. Perhaps it seems unvirtuous to start with the easy stuff—but it's anything but, because you're using the easy stuff to get you to the hard.

Chapter Summary

- Many writers struggle with beginning a new project, or even a new day's writing session.
- Reluctance to start can be unintentional (procrastination) or intentional (believing oneself not yet ready to write).
- Procrastination can be managed with some attention to its psychology, and by manipulating expectancy, value, discounting, and delay.
- Writing can begin before all data and analyses are in hand, and even before you know what your paper will say.
- "Early writing" (writing throughout project design and execution) avoids struggles with starting, makes writing easier, and lets writing help you discover ways to improve the science you're executing.

Exercises

1. For a writing project you have underway, and considering TMT theory, write down some specific actions you could take to increase the attractiveness of writing. Include at least one action each that manipulates expectancy, value, discounting, and delay.
2. Choose an analysis that you plan to report, but for which you have not completed gathering or processing data. Invent some "simulated data" incorporating the pattern you think is most likely, and make a mockup figure or table displaying this "result."

SIX

Momentum

Now that you've started writing (chapter 5), the next challenge is to make sure you keep going: that you build and maintain writing momentum. You probably aspired to a career in science because you liked puzzling out proofs, making chemicals react, tracking wolves through the woods by moonlight, or working with students in the classroom or lab—but most likely you've already discovered that you spend more time writing than doing any of those things. In fact, the volume of writing involved in a typical natural-science career is startlingly large. As a rough calculation, my typical year includes:

- four journal papers, average five thousand words each
- two grant proposals, two thousand and nine thousand words
- twenty-four peer reviews, average 1,200 words each
- one technical report, three thousand words
- six administrative documents, average two thousand words each

My list adds up to about seventy-five thousand words, which is about three hundred double-spaced manuscript pages. All of this must be drafted and then revised repeatedly in response to self, friendly, and peer reviews (chapters 21–23). And, of course, I can't write full time: I have teaching and administrative duties, and I need to spend time actually doing the science I write about!

The point of this calculation is not to brag; if anything, my productivity is low for my peer group. It isn't to scare you, either; if I wanted to do that, I'd tell you about academic administration! The point is to demonstrate that a career in science requires a substantial and sustained pace of writing. This is true even if your job and career stage lead you to a mix of writing activities quite different from mine. (For instance, someone working for government might write fewer journal papers but far more technical reports, while someone at an earlier career stage will write far fewer peer reviews.) To sustain a productive writing pace, you have to maintain momentum even while balancing the rest of your job (and life) duties; you have to avoid distraction and cope with the inevitable "writer's block." This may sound daunting, but if I can do it, you can too. Fortunately, with conscious attention to your writing behavior you can make it much easier, day after day, to stick with the writing you've started.

Is Continuing Different from Starting?

Chapter 5 dealt with sitting down to a blank page to start a new writing project. Is continuing to write really anything different? Seen one way, continuing requires many, many decisions to start again: to begin the day's writing session, a new section, a new paragraph, a new sentence. Any reason for not starting can rear up again as a reason for not continuing. Therefore, lessons from our discussion of starting can certainly be applied to continuing as well.

However, this is not the end of the story. For many writers, starting and continuing seem to present different challenges. Some writers struggle to get started but find smooth sailing after that, while others begin easily but struggle to maintain momentum. There are also techniques that help more with momentum than they do with starting a new project. So: on to momentum.

Discipline

For the majority of scientific writers (chapter 2), there is no way to produce a lot of writing without spending a lot of time writing (Silvia 2007). It takes discipline to spend time writing, even if you'd really rather be doing something else. The novelist J. G. Ballard put it well: "Unless you're disciplined, all you end up with is a lot of empty wine bottles. All through my career I've written 1,000 words a day—even if I've got a hangover. You've got to discipline yourself if you're professional. There's no other way" (Ballard 2003, afterword, 5).

So how do you discipline yourself? Here are some useful techniques:

- **Writing quotas**. The simplest approach to discipline is to set a daily quota, either for output (words produced or revised) or input (hours spent writing). At first, this might seem unhelpful: if you aren't good at enforcing writing discipline, why should you be any better at enforcing a quota? But it works, because a resolution to write "enough" is fuzzy and easily gamed, while a quantitative quota is unambiguous. (There are online tools you can use to manage a quota, such as http://www.750words.com).

  Output and input quotas each have advantages and disadvantages. You might consider an output quota dangerous because it values all words equally, and thus encourages you to write a lot of low-quality material. The most common issue writers have, however, isn't producing writing of high quality as much as producing *any writing at all*. Getting lots of mediocre words down isn't a bad thing, because you can revise them later.

  Word quotas are problematic for a different reason: they use a currency that's natural for the production of text but doesn't easily accommodate production of figures, checking of references, and other time-consuming elements of scientific writing. Input (time) quotas adapt much more easily to the diversity of writing tasks, but they have a problem of their own. What does it mean to "write for an hour"? It certainly means you're at your computer (or legal pad, or whatever). It should mean you're not checking e-mail, playing games, or surfing the web, even intermittently. But what about time spent staring at the ceiling? That's occasionally just what's needed, but if it goes on for long it meets your quota while generating nothing to put on your curriculum vitae. Input quotas arguably track the wrong thing, because rewards for writing accrue for output. There is no single right answer here: you will need to discover whether an output or input quota helps you most to sustain writing discipline.

- **Scheduling**. It's a short step from a time quota to a set writing schedule. Time quotas can falter when the day's writing session gets bumped back by things that seem urgent (even if they aren't). This displacement can leave either too little time to meet your quota or only low-quality time when you're sleepy or burnt-out. Scheduling writing removes moment-to-moment choices about the importance of writing.

It also encourages routine (at *this* time of day, every day, what you do is write—there's no choice of distractions to be made). An interesting variation is the idea of scheduling joint writing sessions with colleagues, which can have the virtue of reinforcing commitment. Writers in many cities have organized "Shut Up and Write" meetups, or you can try virtual meetups instead (e.g., http://suwtuesdays.wordpress.com). Silvia's (2007) excellent book *How to Write a Lot* discusses the virtues of (and techniques for) scheduling in much more detail.

- **Timing**. If you're going to schedule your writing time—or even if you're not—consider writing early in your day. Some studies suggest that willingness to tackle undesirable or difficult tasks behaves like a limited resource that becomes depleted by use (Alquist and Baumeister 2012), and that even very different kinds of tasks draw on the same limited reservoir of willpower (Martin Ginis and Bray 2010). The idea, then, is that conducting writing sessions early gives them priority, allocating willpower to writing over other tasks such as grading, paperwork, exercise, or laundry. This strategy complements, but certainly shouldn't replace, others that try to reduce the need for willpower in the first place (such as scheduling).

- **Environment**. Distractions are the bane of many a writer. (See also chapter 5.) Time quotas mean nothing if they count minutes or hours during which you're actually doing other things, and word quotas can be impossible to reach if you can't keep the focus on writing. Because a conscious decision to resist distraction works only as long as it remains conscious, it's more effective to shape your environment to remove distractions—or at least to remove cues that inform you of the availability of those distractions. A nearby smartphone, a solitaire icon on your screen, a window overlooking a basketball court: none of these things are good ideas for distractible writers. Paying close attention to your behavior for a while (chapter 4) can help you figure out which distractions you're most vulnerable to, and thus which choices or cues to purge from your writing environment.

- **More exotic commitment devices**. Each technique above is what economists and psychologists call a "commitment device": an arrangement or behavior you adopt with the idea that it will help you execute a plan of action otherwise threatened by lack of discipline (Bryan et al. 2010). Commitment devices work by making alternatives to your plan less attractive: difficult, embarrassing, expensive, or what have you. (Note the close relationship to TMT models of procrastination; chapter 5.) Removing games from your writing computer, for example, makes the choice you want to avoid require time-consuming reinstallation of software and also conscious awareness that you're rescinding a previous decision to write. Similarly, you can use an add-on to your browser to block websites that you know distract you (LeechBlock for Firefox and StayFocusd for Chrome are two examples). Writing quotas and schedules draw your attention to lack of discipline, especially if you raise the stakes by publicizing your commitment to them. Well-designed commitment devices can be very effective (Boice 1990, Bryan et al. 2010).

There is really no limit to the creativity with which one can design a commitment device for writing, and many a writer has come up with a more exotic device that works for their own psychology. You can start a weekly writing group with a rule that each participant brings two thousand words of new writing to each meeting, with alternatives bringing shame. You can have a friend hide your car keys, smartphone, or Scotch bottle until you've met a writing quota, the alternative here being time-consuming searching. You can publicly post your writing accomplishment (or lack thereof) each day on Facebook. You can enter a formal contract with financial penalties for missing writing targets. (Payment of the penalty can be automated, for instance via http://www.stickk.com; such contracts may be especially effective if the penalties go to an organization with whose goals you disagree.) You could even emulate the American poet James Whitcomb Riley, who had his friends leave him naked in a hotel room so he'd write rather than drink (Hendrickson 1994)[1].

Binge and Snack Writing

You certainly need to write a lot, but do you need to write a lot *at once*? In other words, is it better to work in occasional long writing sessions or in frequent short ones? Much ink has been spilled on this, with each op-

---

[1] Versions of this writing-naked story (minus the drinking) are also told about Douglas Adams, Sherwood Anderson, Agatha Christie, Harlan Ellison, Alan Greenspan, Ernest Hemingway, Terry Jones, Jean-Paul Marat, and Forrest McDonald, to name a few. Either this commitment device is much more common than one might think, or (more likely) this kind of story is just too much fun to disbelieve.

tion getting a catchy but disparaging name: "binge writing" for the former and "snack writing" for the latter.

There are powerful arguments in each direction. Many scientific writers think that frequent short writing sessions aren't likely to accomplish much, because it takes a while to find your place in the writing and re-establish momentum. This is probably true for very short sessions (less than about half an hour), but some empirical studies (e.g., Boice 2000, his chapter 11) suggest that writers who commit themselves to regular (at least daily) sessions of thirty to ninety minutes are in most cases remarkably productive. It's not clear, however, to what extent any positive effects stem from frequent breaks (staying fresh), short inter-bout intervals (reducing the costs of reestablishing momentum), or simply the large amount of total writing time that accumulates from many short sessions—as long as they really are many, and aren't too short. Boice (2000) provides a longer argument in favor of snack writing. Binge writing, by contrast, offers the advantage of time for detailed thought to tackle a complex task (and scientific writing is full of complex tasks requiring detailed thought). Its biggest pitfall is that scientists who insist on long writing sessions (say, more than three hours) usually have trouble finding available blocks of time. It could be true that most scientific writers would be more productive in a weekly seven-hour session than in seven daily one-hour ones—but those seven-hour sessions, in practice, rarely happen.

As usual, when there are two disparagingly named extremes, the best strategy is probably somewhere in the middle. If you can schedule long writing sessions (with appropriate breaks for physical and mental refreshment) and actually make them happen, that's great—but adding snack-length sessions between them can only help. If frequent short sessions are more realistic for you, then write in those. Just make sure they're very frequent, avoid making them too short, and extend the occasional one to binge length if you possibly can. You may need to experiment to find the mix of longer and shorter sessions that works most productively for you.

It's also worth thinking about the allocation of tasks between longer and shorter sessions. Some writing tasks really are better suited for longer periods of attention. For instance, I find that it usually takes complex and holistic thinking to tackle story-finding (outlining and related techniques; chapter 7) or to draft the Introduction and Discussion sections, and I have trouble making much headway on these tasks in thirty minutes. Shorter sessions work well for me to write the Methods or Results sections, draft graphics and tables, or make any but the most major revisions. You'll discover a good strategy for your own writing sessions if you keep track of what works best for you.

Interruptions, Intermissions, and Momentum

Interruptions in writing sessions, and intermissions between one writing session and the next, can pose a particular problem for maintaining momentum. Returning one's attention to the manuscript after an interruption or intermission takes time and some psychological reorientation. Worse, distractions and writer's block are particularly likely to lurk near the beginning of a session.

One obvious way to maintain momentum, then, is to reduce the number of unplanned interruptions. It's certainly wise to choose a writing environment with that in mind: a private office or carrel instead of a shared one, a phone with its ringer turned off, and so on. It's unrealistic, though, to think you'll never be interrupted. Life does that, so you need to be able to restart. All the advice of this chapter and the previous can help, but there's one additional technique of special value: leaving a loose thread to pick up. When you're interrupted during writing, your first instinct is probably to tell your interrupter, "Hold on a moment while I finish this thought." Don't. Quickly jot down just enough misspelled, point-form notes so that you don't entirely *lose* the thought, but leave the sentence or paragraph deliberately unfinished. You can then rapidly polish the unfinished text when you return. Doing so will get you back into writing and restore the momentum you had before interruption.

Planned intermissions between writing sessions pose similar hazards to momentum; so do the planned breaks you use to refresh yourself during long sessions. (Even very short breaks can increase performance at a sustained task; e.g., Ariga and Lleras 2011.) You can use an unfinished-text strategy to maintain momentum through planned breaks just as you can for unplanned interruptions. No matter how tempting it may be, don't conclude a session of any length by tidily wrapping up a paragraph or a section. Instead, write the first sentence of the next paragraph, or jot outline notes for the upcoming section. This leaves a loose thread for you to pick up when you return to writing. Let that loose thread pull you up to speed.

Storming the Beach

So far I've talked about momentum largely with respect to whether, at a given moment, you are writing or not. But there's more to momentum, once we consider what it is you're actually writing. Kurt Vonnegut, in his autobiographical novel *Timequake*, described two kinds of writers: "swoopers" and "bashers":

> Swoopers write a story quickly, higgledy-piggledy, crinkum-crankum, any which way. Then they go over it again painstakingly, fixing everything that . . . doesn't work. Bashers go one sentence at a time, getting it exactly right before they go on to the next one. When they're done, they're done. (Vonnegut 1997, 137)

Vonnegut was brilliant and knew a lot about writing, but his analysis is incomplete in a very important way. There are, in fact, *three* kinds of writers. Many are indeed swoopers. (I try to be one.) A very few are indeed bashers. (Vonnegut himself rewrote every page until it was pretty much ready for typesetting, before going on to the next.) But he misses a category of writers we can call "draggers." Draggers think of themselves as bashers, but they're wrong. In fact, bashing doesn't work for them: their attempts become endless toil as they drag themselves through writing weighed down by a premature search for perfection. Draggers agonize over perfect phrasing—only to find when they write the next page that they need to go back and rewrite that perfect phrasing anyway. They struggle to write the ideal first paragraph for their Discussion, handicapped by not yet knowing what last paragraph they're setting up. They labor for hours to perfect the format of a figure, but upon making the next two figures realize the first one isn't needed. In the end, draggers would write more easily and more productively if they swooped—but they don't. The fantasy and mystery novelist Matt Hughes puts the case for swooping this way:

Writing a first draft is like hitting the beach on D-Day. You don't stop to tend the wounded or mourn the dead. If you don't get off the beach, you'll die there . . . The point of the first draft is not to get it right, but to get it written. Don't go back and rewrite the first chapter until you've finished the last. Get off the beach. Otherwise, you may never get past page twenty. (Hughes 2011)

If you think of yourself as a basher, be skeptical. It's possible that you really are one, of course; like California condors and good disco songs, bashers are rare, but they do exist[2]. But it's much more likely that you're dragging, when you'd do better to swoop—to storm the beach.

Storming the beach means charging ahead full-speed, although not necessarily in a straight line. It means getting a complete draft committed to paper, no matter how ugly it might be. Actually, the ugliness of a swooper's first draft is completely unimportant: it isn't a *product*, and its quality shouldn't be evaluated as if it were. The first draft is part of the *process* of writing. Nobody but you need see it. If it isn't a masterpiece, you're in good company; most people's aren't. Getting that first draft down also means resolving not to worry (yet) about details. If you aren't sure quite what word to use, don't wait to think about it; insert one that's close and move on. If you like, mark it with a symbol like "???" or a big blank ("_____") so it's easy to find later. If you need a citation and don't have one handy, don't leave your draft to search the literature; just fill in "(???citation???)" and move on. If you're not sure things are in the right order, flag that with "(???reorder???)" and keep writing. If you feel yourself getting stuck for any reason at all, don't stop; move on up the beach.

If you're a dragger, a change to storming the beach might feel unproductive, or at least unvirtuous, because you're leaving all the hard parts for later. In a way, you are, but this is not a problem. At the very worst, finding that right word will be just as difficult later as it would have been when you skipped over it—but it won't be harder. Usually, though, you'll find the "hard" part has become easier since you left it. Sometimes this happens because inspiration has struck in the meantime: you've bought your subconscious time to bail you out while your conscious efforts are

---

[2] Only twenty-two California condors remained in 1987 when all were captured for breeding programs. As of January 2015, the world population was 425 birds, 219 of them in the wild. Good disco songs are much rarer, but Gloria Gaynor's "I Will Survive" won the only Grammy Award ever given for Best Disco Recording, and if a disco song can rock, "I Will Survive" does.

on other things. Sometimes it happens because in writing other parts of the manuscript you've learned things that shed light on the material you were struggling with. Finally, quite often you'll end up excising the troublesome bit anyway, and you won't have to fix it at all. Three ways to win!

"Writer's Block"

Much has been said about writer's block. A writer in this state stares at the page, unsure where to begin or what to say next, for hours, days, weeks, or worse. Writing seems impossible, with only dead ends in sight. The blocked writer may even feel as though they will never write productively again. If you've felt this way, you're in excellent company, alongside such brilliant authors as Maya Angelou, Neil Gaiman, and Barbara Kingsolver. Each of them has felt blocked, but wrote anyway, and you can too.

The key to understanding and coping with writer's block is to understand that it's poorly named. A "block" sounds like an obstacle imposed on you from outside: a concrete barrier across the roadway. But really, writer's block comes from inside the writer. It's the *perception* of obstacles, or at least the failure to see ways around them, that holds a blocked writer back. I point this out not to place blame for the blockage, but rather to help you find the tools to escape it—tools that lie within your own behavior.

It's not hard to find long lists of tips for overcoming writer's block. All the effective ones boil down to this: you feel "blocked" because something in your thinking or behavior is getting in the way of writing the next sentence. The only fix is—wait for it—to write that next sentence anyway, either by meeting the blocking behavior head-on and overpowering it, or by sneakily detouring around it. Fight behavior with behavior. Here are some ways to do that:

- **Lower your standards**. The poet William Stafford said, "I think there are never mornings that anybody 'can't write.' I think that anybody could write if he would have standards as low as mine" (Stafford 1978, 104). If you find yourself unwilling to write because you're imagining readers, reviewers, or even yourself wincing at your terrible prose, you are being blocked by your Inner Critic. Tell that Critic to get lost.

Their job is in revision, and the first draft is none of their business. Hold your nose and write something terrible. If necessary, for ten minutes write something *deliberately* terrible (and have some fun with it). Then move on.

- **Divide and conquer.** You may feel overwhelmed by the task ahead, not sure where to begin. This happens most often at the beginning of a manuscript or section, although even a new paragraph can cause trouble of this sort. You've set yourself goals that are too far out of sight; smaller, more concrete ones will seem more reachable. Put the end of the manuscript (or even of the paragraph) out of your mind, and think instead about the value of getting just the next couple of sentences down. Or take ten minutes to jot an outline of the section or the paragraph, to make that one impossible step into a series of smaller, easier ones.

- **Write *two* versions of the passage you're stuck on.** This one may sound stupid. Surely, if writing one version has you blocked, writing *two* can only be worse? But many writers become paralyzed by the conviction that there's a single correct way of writing a passage, and they haven't found it. There's never just one way of writing anything, and deliberately choosing to write (or at least outline) two versions breaks that paralysis. Don't spend hours on this: aim for ten minutes, and then pick one version to continue. Actually, you needn't even stick with it that long: if you decide to abandon one version after a few moments because the other one is better, that's perfect—you're unblocked.

- **Change your environment**. Get up and go to a new place, and write in a new way. Your new place shouldn't be more than about ten minutes away (because getting there shouldn't become a distraction), but it should be somewhere with different sight, sound, touch, and smell cues. Don't take your usual laptop with you. Take a pen and notebook to the coffeehouse down the street, or to a bench at a bus stop. When you get there, immediately start writing something, no matter how bad it might seem.

- **Talk it out**. For some writers, there's something about the appearance of words on a page that formalizes composition and makes it tempting to expect impossibly quick perfection. So take ten minutes and speak the passage you're working on instead of writing it. Talk to a friend, real or imaginary, or to yourself. Record yourself, and after ten minutes, play back what you've said. Then, no matter how bad it seems, stitch a few sentences into your manuscript to give you a new place to keep going from.

- **Freewrite**. Set a timer for ten minutes, and until it goes, write anything that comes into your head. Don't worry if it's misspelled, ungrammatical, irrelevant, or even gibberish; just don't stop until the timer goes off. At worst, you'll merely get your brain thinking that it's writing (which it is!), but you'll probably find that your freewriting contains some useful sentences that can kickstart the passage you're stuck on.

- **Skip ahead**. Find a piece of the writing project that's straightforward but important: formatting references, resizing table columns, revising a Methods section. Work on this other task for ten minutes, then return to where you left off—with the taste of success displacing that of failure.

- **Back up**. Sometimes the feeling of being stuck isn't really about the passage that comes next; instead, it's about the passage you just wrote. Your subconscious may be telling you that the argument isn't quite logical, that the material is in the wrong order, or that you aren't completely confident in your interpretation of the data. Back up a few lines or a paragraph and try something different in organization, logic, or interpretation. Don't try to rewrite the whole paper from the beginning. Remember, you're storming the beach; don't give up all the ground you've gained, just try to find a slightly different path past the barbed wire.

- **Take a break**. If nothing else works, then take a shower, take a walk, change your clothes, or have a snack—something to break the feedback cycle in which feeling stuck makes you stressed and thus *more* stuck. But make it a short break—ten minutes—and resolve that when you return to your desk you'll immediately write something (good, bad, or freewriting). The short duration and return to writing are critical: otherwise, you're not taking a break, you're succumbing to distraction!

Your sharp eye has no doubt discerned two themes running through these ideas: all involve deliberate changes in your behavior, and all involve only short interruptions in your writing. You need to change your behavior because that's where the block is coming from; just waiting for it to lift would work only if writer's block were an external force. And while you may need to interrupt writing in order to change behavior, the interruptions must be short. That's because in the end, there is no other way to get past writer's block than to just write. You're a professional; that's what professionals do.

Chapter Summary

- A career in science requires a substantial and sustained pace of writing.
- Techniques for sustaining discipline at writing include quotas, scheduling, writing at productive times, setting up a distraction-free writing environment, and many other "commitment devices."
- "Binge" and "snack" writing can each be unhelpful if used exclusively. However, frequent short writing sessions can be surprisingly productive.
- For most writers, "swooping" (rapid production of a first draft, even one of low quality) is far better than "bashing" (revising and polishing as you write).
- Most writers experience "writer's block." Effective ways to overcome it involve deliberate changes to writing behavior, but only temporary interruptions in writing.

Exercises

1. For one week, and for any current writing project, try to meet a 750 words per day writing quota. Record your progress through the week. If you met the quota, what did you have to do in order to meet it? If you did not, what behavior or competing demands prevented you?
2. For one (separate) week, set a writing schedule of at least one forty-five-minute session per day (twice this, if you can). Record your progress through the week. If you were able to stick to this schedule, how did it affect your writing productivity? If you were not, what behavior or competing demands prevented you?

Part III

|||||||||||||||||||||||||||

Content and Structure

Let's presume that you're ready, willing, and able to put fingers to keyboard and write a scientific paper. In Part III, I'll focus on the journal paper, because that's the most common and important kind of writing the early career scientist will grapple with. Later, in chapter 25, I'll turn to other scientific-writing forms. It's now time to ask what content you will put into your paper, and what kind of structure you will use to arrange it.

Getting content right—what I call "finding your story"—is a major part of your writing battle. Fortunately, there are many techniques that can help turn a mountain of data into a compelling story for your readers. The right content, though, isn't enough; you must also present that content in a way that makes it easy for your readers to absorb. Decisions about presentation aren't made in a vacuum: over the last several hundred years we have developed a standard system for structuring and organizing most types of scientific papers. In addition, individual disciplines have conventions that you, as a writer, should learn and respect. These standards and conventions are tremendously helpful because they take advantage of a wealth of experience in how crystal-clear communication is best achieved, and because they align your presentation with what readers expect.

This section doesn't attempt exhaustive coverage of structure and format in scientific writing. Instead, I try to offer general principles that can be applied to making decisions about details. When this isn't enough, every writer should be familiar with reference books that are more exhaustive; I mention some excellent ones in the chapters that follow.

SEVEN

||||||||||||||||||||||||||||||||||||||||||||||||||||||||||||||||||||||||||||||||||||||||||||||||||||||||||||||||||||||||||||||||||||||||||

Finding and Telling Your Story

The very first step in writing any scientific paper is to answer a simple question: what, exactly, will it be about? That is, what's your story?

Your "Story"

That your paper will be about your results—data from your experiments, proof of a theorem, observations with a new telescope—is obvious, but for two reasons, it's also unhelpful. First, it's rare for your results to consist of a single analysis from a single experiment that worked as you planned it and yielded the outcome you expected. Instead, you'll normally sit down to write with a mountain of data from multiple experiments and observations, some obviously critical to your research question, some obviously tangential, and an awful lot somewhere in between. Some results may be unexpected and some apparently conflicting. For each dataset you may have several alternative analyses. You may even find yourself piecing together datasets collected for different reasons by different people long before your own involvement, to test a hypothesis that was never considered when the data were collected. In other words, it may be obvious that your paper is about your results—but *which* results, and why?

Second, saying that your paper is about your results places the focus on the writer ("*your* results"), which is not where it should be. Good writing is oriented to the needs of the reader. You should ask not "what should I write about?" but instead "what does my reader need to hear about?" The distinction is subtle but important, and it leads you to identifying the story you have to tell.

What does it mean for your paper to tell a story? Successful fiction or drama sets up and then resolves some interesting question in a reader's mind, by exposing compelling characters to a well-defined plot (well, except for *Waiting for Godot*). A scientific paper does the same. It has characters: the rocks, chemicals, equations, or other entities that you studied. It has a plot: the methods you applied to your characters and the results you obtained from them. Most importantly, it raises and answers an interesting question.

The central question and its answer are the purest distillation of your story. A clear central question gives your paper a single, obvious direction. Every element of your paper can then work together to draw the reader inescapably to your question's answer. You should be able to state question and answer in a sentence or two—perhaps not in a way useful for a reader, who would need definitions, context, and so on; but in one that defines the story for you. Whenever you are not sure whether a dataset, analysis, figure, table, or anything else belongs in the manuscript, referring to this two-sentence mini-summary should give you the answer: would including it help tell the story or distract from it?

For a concrete example (one we'll return to repeatedly over the next several chapters), imagine that you're an astronomer interested in the formation of massive stars (Box 7.1). You have used the new Atacama

---

Box 7.1 Star formation

Briefly, stars form as clouds of interstellar dust and gas collapse under their own gravity. As a dense core (a protostar) forms, gravitational potential energy is converted to heat. The heated material emits radiation, supplying an outward force (radiation pressure) that increasingly opposes the gravitational pull on remaining matter in the cloud. Matter accretes to the protostar until it has become large and hot enough for radiation pressure to balance gravitational pull.

This process is well understood for small stars, but not for larger ones (more than about ten times the mass of our sun). The simplest models suggest that radiation pressure should become too strong for further accretion before the protostar can reach such masses. One possibility is that radiation doesn't escape equally in all directions, but rather in concentrated jets that clear away material in one direction but allow accretion from others (Banerjee and Pudritz 2007). Another is that massive stars form at the center of star clusters, where accretion can be driven by the combined gravity of many protostars (Bonnell and Bate 2006). The question is interesting because massive stars are rare, only about 0.2% of all newly formed stars (Chabrier 2003), but we don't know why.

The rival models of massive-star formation make different predictions about the appearance of massive protostars and their distribution in space, which are testable with sufficient observational data. Fortunately, gas clouds containing protostars are fairly common, with at least 6,000 large clouds and many smaller ones in our galaxy alone (Sanders et al. 1985). Regions of massive-star formation close enough for detailed imaging include the Orion, Eagle, and Carina Nebulae (1,300, 7,000, and 8,000 light years from Earth).

Large Millimeter/Submillimeter Array (ALMA; a radiotelescope complex in Chile) to image the Eagle, Orion, and Carina Nebulae, where new massive stars are being formed. ALMA is more sensitive and has higher resolution than any other millimeter-wave telescope, and you have reams of imagery, spectral data, and so on. But what story will you tell? Here are two-sentence summaries of some ways you might present your ALMA data. These are framed informally, because they aren't intended for anyone but the writer.

Very bad:

ALMA provided the most detailed images ever made of the Orion, Eagle, and Carina Nebulae. I explain how ALMA works, and show some of the images.

This mini-summary focuses on the research done, rather than the question answered. It makes no reference to star formation (or to any question at all). It forces the reader to do the job of analyzing and interpreting the data, but most readers will simply move on.

There is, of course, always more than one story that could be told with a set of data. If the paper was intended to report on ALMA's capabilities, rather than to test models of star formation, this mini-summary wouldn't be as bad.

Bad:

I present images of star-forming regions in the Orion, Eagle, and Carina nebulae, taken at 8 wavelengths from 0.4 to 10 mm. They show many protostars, some in groups.

This summary mentions star formation, but asks no specific question about it. It defines methods and offers data that are presumably relevant, but the job of figuring out how is left to the reader.

Better, but not good:

I outline what we know about star formation and stellar evolution. I present ALMA images of massive protostars at various stages of stellar evolution.

Here the writer makes some attempt to establish the relevance of the data, but "star formation and stellar evolution" is far too broad for a single paper, and no specific question is offered. The images presumably tell the reader something about massive-star formation, but it's not clear exactly what.

Very good:

If massive stars form with gravitational assists from their neighbors, then massive protostars should always appear among other protostars. However, solitary massive protostars are common in ALMA images of the Orion, Eagle, and Carina nebulae.

This mini-summary has a clear central question (do massive protostars form via combined gravity of neighbors?), and offers an answer (no). The question drives the function of each part of the manuscript: the Introduction will set up the cluster-gravity hypothesis and specify what's needed to test it; the Methods will explain how we can measure protostar masses and distinguish between solitary protostars and sets of neighbors; the Results will tabulate frequencies of solitary and neighbored protostars; and the Discussion will interpret the Results as a test of the cluster-assist hypothesis.

Perhaps the case for having a clear story is so obvious that it's hard to picture yourself sitting down to write without one. However, it's remarkable how often I see draft manuscripts that betray their authors' uncertainty about the topic. Among the symptoms: a long and convoluted title, or a short but vague one; data presented but never analyzed; figures or tables that don't relate to any hypothesis raised in the Introduction or that don't contribute to the Discussion; or topics appearing in the Discussion that weren't broached earlier in the manuscript. The writers of such manuscripts have omitted a critical step in their work: they have not found their stories.

Finding and Planning Your Story

Finding and planning your story means accomplishing three things. First, you must identify your central question and its answer (the two-sentence mini-summary from the last section). Write this down, even though no reader will ever see it; the writing will force you to be explicit. Second, you must decide which information, data, analyses, and interpretation belong in the paper, and which are better reserved for another manuscript or abandoned to a dusty row of old notebooks. Third, once you've identified the content, you must decide the order in which to present it. Of these, the first point is finding a story to tell, and the second and third are planning the best way to tell it.

Because the path from concept through execution to data analysis and interpretation is rarely simple, finding your story uses hindsight as much as foresight. Writers who don't realize this often cling to reporting everything they did, in the order they did it, including experiments that turned out to be blind alleys and observations that seemed relevant at the proposal stage but became immaterial to the conclusions eventually drawn. Experienced writers, in contrast, take full advantage of hindsight and work to determine the story they want the reader to hear (whether or not it's the one they had in mind when they began the work), and what information the reader needs to understand that story.

Do not hope to find your story by sitting down at your keyboard and beginning to type the first line of your Introduction. Instead, begin by brainstorming possible content and then selecting and organizing the content that defines your story and tells it effectively. There are many techniques for finding and planning your story; what follows is a toolbox from which you can select. Many writers apply more than one of these techniques to each writing project.

Figure 7.1. My wordstack for this chapter.

cohesive story. . . . "thesis"
?titles - shortest summary of story
outline
story about how I thought I didn't do outlines
when to outline – when story is ready; vs. to find the story
head and subheads as coarse outline
topic sentences as detailed outline
what goes in and doesn't, and what order

not everything you did
not in the order you did it

concept map: non-linear
intermediate step to outline
despite HTML, basic form still linear
wordstack/idea pile

first step
accumulate pre-writing

?Cahill – "pitch"
fail to consider story: leads to overlong, poorly organized MS
IMRaD structure
online supplements
"retroactive storytelling"
avoid lock-in
simple clear direction
selling the story

not "I was interested in"
not "no studies have examined"
not "increase our understanding of"

contrast "writing backwards" (Magnusson 1996)
figure shuffling

Wordstacks. A wordstack (Figure 7.1) is an unsorted list of points you think might be useful ingredients in your manuscript. Each point can be a single word or short phrase indicating a relevant fact, idea, or topic, or can be a roughly sketched graphic. Your wordstack might have some hierarchical structure (some points having subpoints) if this is immediately obvious, but don't force it.

I illustrate wordstacks and the next two techniques with examples for this chapter because doing so makes clear the relationship between the wordstack and the finished product. Notice that it isn't a very close one. A wordstack is a venue for brainstorming, and it's more important to get ideas down than to worry about their being fully developed, related to each other, or in logical order. For the same reason, don't worry if you're not sure an entry really belongs: note your skepticism (I use question marks), but don't remove the entry. The point of a wordstack is not to organize your material or fully define your story, but rather to display the raw material from which you can draw. In moving from wordstack to finished chapter, I included one point I'd marked as questionable (pitch) but left out another (titles). I also left out material I hadn't initially questioned (a story about my outlining habits) and included material not in the wordstack (the story summary). All this is normal: the wordstack is a tool, not a product.

Concept maps. A concept map (Figure 7.2) is a tool for exploring relationships among concepts (Novak and Cañas 2008). It consists of a set of nodes connected by lines. Each node is a concept: a word or phrase denoting an idea, a thing, or a property of one of these. Each line connects two nodes and is labelled to indicate the relationship between those concepts. Typically, most nodes will be nouns or adjectives, while most line labels will be verbs.

To construct a concept map, first identify (provisionally) the most general concept involved in the topic of your manuscript. For this chapter, I chose "clarity"; for our astronomy example, one might choose "star formation." Then add concepts from your wordstack or others that suggest themselves as you build the concept map. As you add each concept, think about which others it relates to, and how; indicate these relationships with labelled connecting lines. I added "appropriate content" to my concept map, and indicated that clarity depends on determining this; I later broke "appropriate content" down by indicating that it consists of "content items" that are "few" and "ordered." Of course, with sufficient imagination one could connect every concept to every other one, so part of the job is to choose the relationships that you want your reader to think about. Place more specific, narrowly defined concepts

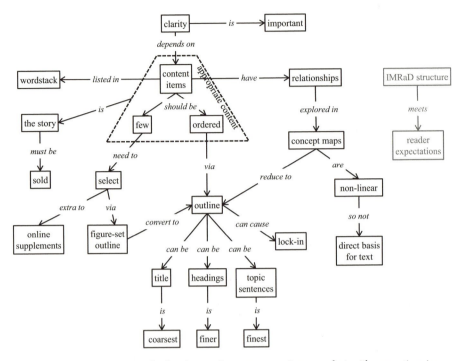

Figure 7.2. My concept map for this chapter. Concepts appear in roman font, with connections in italic. Concepts in gray failed to connect to the core structure, and were omitted from the chapter.

lower down, so that your concept map is roughly hierarchical. You will likely rearrange your concept map repeatedly as it grows (perhaps using CmapTools software: http://cmap.ihmc.us).

The completed concept map depicts your decisions about the logical relationships among material you might include in your manuscript. It doesn't fully define your story, because it's likely to show too much complexity, but your story is implicit in the main flow of connections among concepts. You may find that your concept map includes nodes or sets of nodes that are not connected to the rest of the map (for example, the gray nodes in Figure 7.2). These suggest material that doesn't belong. Your most important material is likely to appear near the center top of your concept map, with strong connections to other concepts—as it does in my example (the trapezoidal supernode in Figure 7.2).

An important property of concept maps is that they can, and usually do, incorporate branching, loops, and other network structures (as in the three paths from "content items" to "outline" in Figure 7.2). This is a strength of the technique for initial exploration of your material, but a shortcoming of its use for translating material to text. We've spent several decades experimenting with crosslinked hypertext on the World Wide Web, but a linear sequence of ideas is still the most effective way to communicate complex information. It's the writer's job to determine *which* linear order makes the material clearest for the reader. Converting a concept map into a manuscript, then, requires an intermediate step to linearize it. The best way to do that is to construct an outline.

Outlines. An outline (Figure 7.3) is an ordered list of topics or points that summarizes the intended content of your manuscript. It is the result of selecting and ordering topics from the larger set available in a word-stack, or of linearizing a concept map. An outline has an intended 1:1 mapping onto a completed draft, such that expansion of each outline point, in turn, into a section of text is sufficient to produce the draft.

Figure 7.3. My outline for this chapter.

1. The concept and importance of your "story"
    1.1 what's a "story" (simple, clear direction)
    1.2 importance of story
2. Planning the story
    2.1.1 wordstack
    2.1.2 concept map
    2.1.2.1 use in learning
    2.1.2.2 use in writing
    2.1.3 outlines
    2.1.4 story summary
    2.1.5 subhead outline
    2.1.6 topic-sentence outline
    2.1.7 figure-set outline
    2.1.8 title
3. Hindsight storytelling
4. Revising the outline
5. Online supplements
6. Selling the story

(That this mapping is *intended* recognizes that reorganization during writing is common; see "Revising the outline," below, and notice the differences between the outline in Figure 7.3 and the chapter you are reading.) In addition to identifying and ordering included topics, outlines usually indicate hierarchical organization of subtopics into topics.

While there may be as many ways to outline as there are writers who outline, three particularly useful approaches are the story summary, the subhead outline, and the topic-sentence outline.

- **The story summary.** A story summary consists of answers to the following nine queries about your work and your story:

  1. What is the central question?
  2. Why is this question important?
  3. What data are needed to answer this question?
  4. What methods are used to get those data?
  5. What analysis must be applied for the data to answer the central question?
  6. What data were obtained?
  7. What were the results of the analyses?
  8. How did the analyses answer the central question?
  9. What does this answer tell us about the broader field?

(If you're a theoretician, to apply the story summary to your work, think of "data" defined broadly as the outcome of models.) Think of the nine queries as a form with nine fields to be filled in. Queries 1–3 and 8–9 should be answered with a single sentence each. Queries 4–7 may need a bit more if multiple experiments, datasets, or analyses must be combined to answer the central question, but a few short sentences should suffice. For our star formation example, for instance, query 4 might need three elements:

4a. *Rotational velocities from red/blue shifts of ^{13}C emission line.*
4b. *Protostar masses from plot of radius vs. rotational velocity.*
4c. *Nearest-neighbor distances from annual parallax[1] of each protostar.*

---

[1] Annual parallax (the shift in direction to an object viewed from opposite sides of Earth's orbit) indicates distance from Earth to the object.

When the story summary is complete, you've outlined a whole paper: answers 1–3 for your Introduction; 4–5 for your Methods; 6–7 for your Results, and 8–9 for your Discussion.

A story summary doesn't provide a complete manuscript outline. You will likely expand it later into a subhead or topic-sentence outline. Its value lies in forcing attention on your story and on what material the reader needs to understand it. For example, query 3 doesn't refer to all the data you have gathered; it asks about the data needed to answer your central question (a different and usually much smaller set).

- **The subhead outline**. A subhead outline is made up of phrases or other entries intended for use as headings and subheads in the completed manuscript. For a standard scientific paper, the top-level headings are nearly always Introduction, Methods, Results, and Discussion (chapter 8), although they will differ for other writing forms. Each top-level heading can receive subheads dividing it into logically distinct subtopics, and these can in turn be divided further. The manuscript is then written around the subheads, as blocks of text are inserted under each.

  Think carefully if you find yourself tempted to use more than three levels of headings (such as, for our star-formation example, "*Methods—Protostar neighbor distances—Parallax measurement*"). The function of subheads is to communicate organization to the reader, and if the organization of your paper into major topics is complex enough to require fourth-level subheads, it's probably too complex! Material below third-level heads should usually be simple enough for a reader to be guided through by normal paragraph and sentence structure (chapters 17–18). If you feel the urge to elaborate your outline further, that's fine—but do it by a topic-sentence outline instead.

- **The topic-sentence outline**. A topic-sentence outline has one entry for each intended paragraph of the completed manuscript. Each entry is a complete sentence that expresses the topic of its paragraph (and is thus suitable for use as the first sentence of that paragraph). Topic-sentence outlines are more detailed than subhead outlines, in two ways. First, they are more finely resolved. Second, they specify the material to be written more completely, because they consist of logical statements that summarize points to be made, rather than just naming topics. Returning to our star-formation example, a topic-sentence outline for the third-level head *"Methods—Protostar neighbor distances—Parallax measurement"* might consist of two sentences:

> Parallax of each protostar is measured by change in right ascension and declination[2] between March and September sightings.

> Right ascension and declination are measured relative to distant galaxies, for which parallax is negligible.

As you write, you'll expand each topic sentence into a paragraph giving further detail: in this case, how the measurements are made, which reference galaxies are used, and so on. A topic-sentence outline is complete when it contains enough detail for this expansion to seem straightforward to you. While this is an imprecise standard, it isn't very important to know when your topic-sentence outline is "complete"—because the process of expanding it simply grades into writing of the manuscript itself.

Figure shuffling. Figure shuffling is an alternative to outlining that focuses on data and analyses as the elements that define your story. Before writing, you probably spend lots of time doing exploratory data analysis: plotting relationships between different pairs of variables, summarizing data in tables, running alternative statistical tests, and so on. You may have dozens of rough figures and tables at hand; some belong in your manuscript and some don't. Figure shuffling involves pinning these up on a wall and winnowing and shuffling them to produce a set of reasonable size representing the story you want to tell.

When you're done, every figure and table you've selected should be essential to telling your story. If one isn't, take it out. How many should remain depends, of course, on the work and on conventions in your field, but I think you should be reluctant to include more than about five figures and four tables in a single paper. If you need more, you may not have found your story. You will also have decided on the order in which

---

[2] Right ascension and declination are astronomical coordinates that express the direction from Earth to a viewed object; they are analogous to longitude and latitude.

the figures and tables make the most sense. Shuffling figures without thinking about methods allows you to ignore the order in which you did the work (irrelevant!) and focus on the order that best communicates the results.

Figure shuffling may seem to put the cart before the horse, because it doesn't start by identifying your central question. Instead, it finds the answer, and the question is implicit in that. Figure shuffling can be thought of as a way to "write backwards" (Magnusson 1996). When you write backwards, you begin by identifying your most important conclusion. Then you write your results, including only those necessary to support the conclusion. Figure shuffling accomplishes these two steps. Next, you write the methods necessary to obtain the results, and then the discussion that sets your conclusion in a broader context. Finally, you write the introduction, setting up the central question to which your conclusion is the answer. This technique can be very effective in finding your story, because it divorces the presentation from *what you started out to study* and *what you did* and focuses it instead on *what conclusion the reader should take from the work.*

Wordstacks, concept maps, various kinds of outlines, and figure shuffling can complement each other, but few writers use them all. My own mainstays are wordstacks, subhead outlines, and topic-sentence outlines; constructing those (in that order) usually suffices for me to find and plan my story. Experience will show you the set of techniques that helps you find and plan yours.

A Caution on Leaving Things Out

Leaving things out is critical to finding and telling your story. However, it's very important to think carefully about what it means to present "only those results necessary to support your conclusion." This does *not* mean omitting results that conflict with your conclusion, which, of course, is unethical. It does mean omitting results that are not relevant to your conclusion, or are redundant with others that suffice to support it. These are some of the most important judgements you can make as a scientist, and there is no simple prescription for making them other than experience and careful thought.

Revising the Outline

Finding and planning your story is essential, but danger lurks in outlining, concept mapping, and the like nonetheless. Remember that these are tools, not straitjackets. As you flesh out an outline into a manuscript, you may feel something not working—perhaps a topic appearing early in your outline doesn't seem to fit comfortably there anymore, or a new topic is clamoring for inclusion. Should you stick to your outline, which represents the story that you've planned, or change it?

The answer, of course, is that it depends. On one hand, there's no point using a story-planning technique if you're going to completely ignore the plan you've made. On the other, the thinking you do as you write can change your interpretation of your data, or otherwise alter the story you're telling. Getting locked into the outline fossilizes your thinking and closes off avenues for improvement.

The happy medium is to think of your outline as providing an explicit criterion against which you can measure a potential change. Ask yourself critically: does adding new material to the outline improve the story? If you can explain why it does so, then include it; if not, it's distracting you from your planned story. Does reordering topics or deleting an outline point let you tell the story better? If and only if you're sure that it does, make the change. For example, in writing this chapter from the outline in Figure 7.3, I deleted my intended section 2.1.2.1 on concept maps in learning, after realizing it was irrelevant to my story. I moved section 5 on online supplements into chapter 14, where it fit better. Section 3 on hindsight storytelling turned out, once I'd written most of it, to fit better near the beginning of section 2 on planning the story. In each case, I thought carefully about which organization suited my story better and then made the change. This kind of reluctant willingness to change the outline makes story-planning a dynamic process that continues throughout writing—but a self-guiding one, as the current plan always provides a benchmark for assessing revisions.

Selling Your Story

Finding and planning your story is part of your job, as a writer, to work toward effortless reading. But your job doesn't end there. Remember that your work competes for readers' attention with an ocean of published material. A story that reads effortlessly improves your competitive position, but you also need to tell readers why they should spend their time reading *your* work, rather than somebody else's. You need to sell your story.

You may be uncomfortable with this. We are told from childhood not to brag, and many scientists are introverts at heart. This discomfort is likely responsible for such timid offerings as "I was interested in studying X," "further studies are required to increase our understanding of X," or "no studies have reported results of experiment/observation X." Many conceivable studies could pass these tests and yet be of little interest to readers. It's tempting to think that your good science will speak for itself, and readers will know why it's important. But few will invest the effort to read through a paper unless its importance is established explicitly right up front. So sell you must.

It's crucial to understand that what matters is your central question's importance to your reader, not to you. Perhaps you were motivated to study massive-star formation by your intrinsic fascination with the physics of collapsing gas clouds. That's perfectly fine as a reason to go to your office every day, but it won't get you an audience. Instead, you need to connect the narrow subject of your manuscript to more general issues that people care about. For instance, nobody knows why massive stars are rare, but they play an important role in the universe: only in such stars can fusion produce elements heavier than carbon, and only through their supernova explosions are such elements available for incorporation into smaller stars, planets, and people. You could explain that your study of massive-star formation will help us understand the process that made the universe suitable for life.

There are alternative ways to sell any story, and the approach you choose determines the set of interested readers. This, in turn, determines the journals for which your manuscript is appropriate. Manuscripts with larger sets of interested readers tend to be accepted by higher-impact journals and to accumulate more citations. Among common and effective ways to sell a story are pitches like these:

- "There's a controversy in the literature over issue X, and I present the kind of data needed to resolve it."
- "The fact that we don't know X hinders our efforts to understand issue Y, which is central to a developing subdiscipline."
- "Our lack of understanding of thing X impedes our efforts to solve economic problem Y."
- "We need to know more about thing X because it's a model system widely used to investigate problems in field Y."
- "I have discovered thing X, which suggests a way to make progress toward difficult-to-reach goal Y."

Any manuscript can be pitched in different ways—those above, and more. Some pitches will excite more readers than others, but just as important, a chosen pitch may excite a particular audience or demonstrate a fit with a particular journal. You can think of these different pitches as slightly different definitions of your story, and you can put them to work to best sell what you have to say to the audience you'd like to say it to.

Chapter Summary

- A paper has a story, with "characters" and a "plot," and it raises and answers an interesting question.
- Tools for finding and planning your story include the two-sentence mini-summary, wordstacks, concept maps, figure shuffling, and outlining. Outlines may be story summaries, subhead outlines, or topic-sentence outlines.
- Telling your story isn't enough; you must sell it, too. This means showing how your work solves a problem, or answers a question, that matters to readers.

Exercises

1. For a paper you've recently read, write a mini-summary, a concept map, and a story-summary outline. Can you suggest an alternative way of organizing the same content, and does your alternative tell a better story or a different one?

2. For a writing project you've recently started or plan to start soon, write a mini-summary. Next, make a wordstack, a concept map, and a subhead outline. If you wrote a new mini-summary, would it differ as a result of the story-planning process?

3. For the project outlined in (2)—or if possible, for a classmate's or colleague's—write three different selling pitches, each no more than two sentences.

EIGHT

||||||||||||||||||||||||||||||||||||||||||||||||||||||||||||||||||||||||||||||||||||||||||||||||||||||||||||||||||||||||||||||||||||||||||||||

The Canonical Structure of the Scientific Paper

The canonical structure of the modern scientific paper is familiar to anyone who has even dipped a toe into the literature. It's often labelled "IMRaD," for *Introduction, Methods, Results, and Discussion,* because it includes those elements in that order (with each section containing a rather standard set of components). This canon represents a consensus among scientists about the most effective way to package information, and it's extremely helpful for both the writer and the reader. It hasn't always been this way, though, and it's worth thinking about why.

The Evolution of Canonical Structure

If you browse seventeenth- and eighteenth-century issues of the *Philosophical Transactions,* you'll find that the earliest scientific papers held to no discernible conventions of structure, format, or style. Some were written as letters (right down to salutations and signatures), while others were travelogues, descriptions, or narratives. Few papers had section headings or other formal structuring, and in those that did, sectioning was idiosyncratic. Literature citations might appear in the margins, as footnotes, as endnotes, in the text, or not at all. Figures and tables were rare, often unlabelled and unnumbered, and often printed at the end of a volume rather than with the paper that referred to them.

Reading this early literature can be very entertaining, but it can also be frustrating and bewildering. Each paper takes its own approach, and it's often not obvious what story is being told and how. This probably wasn't fatal to the writers' aspirations to be read, as the literature at the time was manageably sized and someone could fairly easily read everything published on a topic as broad as "chemistry." Through the nineteenth century, however, the professionalization of science pushed writing away from the descriptive, personal style of the earliest publications. At the same time, the rising volume of published work made readers less patient. In the 1830s and 1840s, an organizational style including elements we would now recognize as Introduction, Methods, Results, and Discussion (Harmon and Gross 2007) became common, especially in German chemistry papers, although these were not necessarily separated and labelled as sections. Detailed methodology became particularly important in work by early microbiologists (such as Louis Pasteur) in support of the germ theory of disease and in rejection of spontaneous generation, and so a separate section for Methods became common (Day and Gastel 2006). However, only in the mid-twentieth century did papers take their fully modern form, with the explosive growth of research and publication during World War II and the space and arms races of the 1950s and 1960s. Journals began to insist on standardized structures with separate and labelled Abstract, Introduction, Methods, Results, and Discussion sections, in part to ease the burden on editors, reviewers, and readers.

The canonical structure we now use evolved to allow efficient access by readers to the content of a scientific paper. Our familiar conventions work as a "finding system" (Gross et al. 2002) in which writers meet well-defined reader expectations: someone wanting to know how an experiment was done can proceed directly to the Methods section; someone wanting a summary of the research question can look to the end of the Introduction; and someone wanting to know why the results are important can check the end of the Discussion. This allows readers to access specific information without having to read the whole paper, but it also assists the start-to-finish reader because it presents information in a familiar order designed to draw that reader inexorably through the paper's story to its conclusion. Finally, the headings, figure numbers, and other organizational features signal the reader's place in the story. Not every paper can take the IMRaD structure (chapter 16), but for those that can, the canonical structure is a powerful tool for achieving crystal-clear communication with the reader.

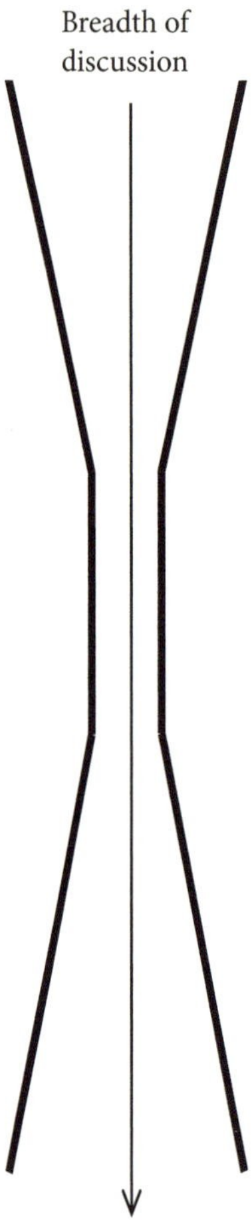

Front Matter
  Title
  Byline and author affiliations
  Keywords, word counts, and other details
Abstract
Introduction
  General context of the work
  Narrower research area and statement of its
    importance
  Identification of a gap or other need for research
  Specific research question meeting the identified need
  Summary of approach to answer the research question
  Announcement of principal findings
Methods
  Materials, species, field sites, mathematical tech-
    niques, etc. used in the research
  Observational/experimental procedures followed
  Methods for analysis of data
Results
  Results of observations, experiments, or modeling in
    text, tables, and/or figures
  Comparisons among results (e.g., observation vs.
    theory or treatment vs. control)
Discussion
  Interpretation of results to answer research questions
  Consideration of possible weaknesses
  Relationship of results to previous literature and
    broader implications of having answered research
    question
  Prospects for future progress
Back Matter
  Acknowledgements
  References cited
  Appendices or online supplements
  Data archives

Breadth of discussion

Figure 8.1 Canonical structure of the modern scientific paper. Sections in black ("IMRaD") are the core elements that tell the paper's story. Sections in *italics* are supporting materials with other functions.

The Canon: "IMRaD" Structure and the Hourglass

The familiar "IMRaD" structure (Figure 8.1) is actually a bit more complicated than its acronym, for three reasons. First, the four major sections are the core of the paper but are surrounded by other elements. The core sections are always preceded by front matter (title, bylines, and other administrative details) and usually by an Abstract; and they are always followed by some back matter (acknowledgements, references, and/or appendices or online supplements). Second, each of the four major sections also has some standardized substructure. Third, there is some minor variation in presentation of the Results and Discussion sections: they are usually (and best) kept separate, but sometimes combined; and the Discussion is sometimes followed by a separate Conclusions section. Nevertheless, IMRaD is the skeleton on which nearly every paper is built.

The IMRaD core has an important property, illustrated by the hourglass shape in Figure 8.1. A well-written paper follows a predictable change in focus: broad attention to the work's context in a major field at the beginning of the Introduction, narrower definition of the central research question at the Introduction's end, narrowest focus on specific techniques and results at the hourglass's middle in the Methods and Results, and broad context again at the end of the Discussion. The broad beginning and ending sell the story (chapter 7) to the largest possible set of readers, while the narrowing through the Introduction defines that story and identifies precisely how the writer will answer the central research question. Keeping the hourglass shape in mind will help enormously in writing effective Introduction and Discussion sections.

Over the next several chapters, I focus on each section in turn. I present them in their order of appearance in the finished manuscript, but remember that few writers will tackle them in this order (chapters 5 and 7).

Chapter Summary

- The IMRaD structure is now standard for most scientific papers, and includes Abstract, Introduction, Methods, Results, and Discussion

- IMRaD functions as a "finding system" for readers, allowing efficient access to content.
- A well-organized paper will have an "hourglass" structure, with focus broad at the beginning, narrowing through the Introduction, and widening again through the Discussion.

NINE

||||||||||||||||||||||||||||||||||||||||||||||||||||||||||||||||||||||||||||||||||||||||||||||||||||||||||||||||||||||||||||||

Front Matter and Abstract

Because the IMRaD core of your paper tells its story, it's tempting to dismiss the rest of the paper—front matter and back matter—as unimportant. This would be a mistake. Together, these additional elements connect your work to the larger literature and the larger scientific community. They provide summary and indexing material so that interested readers can find the work (and its authors) and so that editors can deal with it appropriately in review. Finally, they let you supplement your core story with additional material that can be accessed by others who wish to expand on what you've done. So although front and back matter may be easily written, they should not be treated lightly.

Title

Every paper begins with a title. Its function is advertisement: it invites a potential reader to pick up your paper and read further. Like a pickup line, a title needs to do its work quickly: someone scanning a list of papers may decide whether or not to read yours based on just a few seconds' glance.

To be effective in its advertising function, your title should be brief, clear, and informative. It should communicate your paper's story, or at least its central question; in fact, you could think of your title as the shortest possible summary of your paper. The obvious tension between keeping a title brief and having it summarize your entire paper accounts for the high frequency of "colon titles": longer titles broken by a colon into a more general opening phrase or clause and a following one, usually more tightly focused.

Consider some possible titles for our star-formation paper (chapter 7). First, two very brief but vague options:

Spectroscopic observations of the Eagle, Orion, and Carina Nebulae

Some observations on protostellar masses

These are weak titles. The first emphasizes the paper's methods, rather than its question. The second identifies protostellar mass as an issue but suggests no question about it, and "some observations" is so vague as to be useless to a reader. Neither calls much attention to itself: at best, each clears its throat politely in a noisy bar, and will certainly be going home alone.

Stronger titles put themselves out there, giving a prospective reader a clear indication of the paper's story and even suggesting its importance. For instance:

Protostar distribution and the formation of massive new stars: testing the cluster-assist model

Can patterns of protostar distribution within molecular clouds distinguish between competing models of massive star formation?

Detailed images of protostar neighborhoods do not support the cluster-assist model of massive star formation

Each of these titles indicates the paper's central question, although the first and second give different emphasis to testing a single hypothesis vs. discriminating among competing hypotheses. The third is what's often called an "assertive sentence title," which uses a declarative sentence to state the central question and assert its answer. Such titles are common, but not universally loved. They score very high for clarity and information content, but some readers find them brash and presumptive, exaggerating a claim without supplying evidence or allowing for qualification.

Titles of scientific papers are sometimes funny: for instance, "Escape from the menace of the giant wormholes" (Coleman and Lee 1989) or "The good, the bad, and the cell type–specific roles of hypoxia inducible factor-1 alpha in neurons and astrocytes" (Vangeison et al. 2008). Whether this is a good idea is the subject of some debate. This issue is related to a broader discussion about humor and beauty in scientific writing, which I address in chapter 28.

Byline

A byline, or list of authors with their academic affiliations and addresses, follows the title. The byline has three functions. First, it allows indexing of your paper under your name in databases such as the Web of Science and Google Scholar. Second, it allows readers interested in learning more to contact you directly. Third, it helps establish the authority of your work by indicating your membership in the society of science (your coauthors, your affiliation with a respected institution, and so on).

Writing the byline is mostly a matter of following journal instructions, but two issues may require some thought. When a paper has multiple authors, you must determine the order in which they are listed. This complex issue is treated in detail in chapter 26. You must also decide how to list your name: in my case, I might be Steve Heard, Stephen Heard, Stephen Bruce Heard, Stephen B. Heard, or some other variant. This matters because name variation makes it difficult for someone searching for your publications to find them all. I publish as "Stephen B. Heard," but just once I appeared as "Steve Heard," and most searches for my publications miss that paper. If the searcher wants to assess me for hiring or promotion, is sizing up my lab as a place to do graduate work, or just wants to read and cite my papers, then the name confusion has cost me. You should, therefore, decide on a "publishing name" once, at the beginning of your publishing career, and stick with it. It helps with searches if this is as distinctive as possible, so as a rule you should include middle initials if you have them (and perhaps invent some if you don't). Note that there's no reason you have to publish under your legal name! If you change your name after you begin publishing (via marriage or divorce or for any other reason), you can keep using the older name for publications even while adopting the new name in other spheres. This is not an uncommon practice, and it avoids the risk of your being considered two people by databases and their users.

A new and different solution to the author-tracking problem is the ORCID identifier (Open Researcher and Contributor ID; http://www

.orcid.org), a unique sixteen-digit identifier associated with an individual researcher. If this becomes universally adopted, your ORCID will allow someone to track all your contributions (not just published papers) across platforms, no matter what variant of your name appears on them or how many other researchers share your name. However, ORCID is not yet in universal use, so keeping a distinctive and consistent publishing name remains important.

Other Front Matter

Other front matter will vary with the journal. You may be asked to provide keywords to facilitate assignment of your manuscript to reviewers. (Once these were important for indexing, but with full-text search widely available, this role has faded.) You may be asked to provide word counts, lists of figures and tables, the date of submission, or other information. This kind of front matter facilitates the editorial process, but has nothing to do with telling your story. Follow journal instructions.

Abstract

The Abstract is a recent addition to the canonical structure. Before the 1920s, journal papers rarely carried Abstracts, and their inclusion did not become routine until the 1950s (Harmon and Gross 2007). The Abstract, like the title, has an advertising function: a potential reader intrigued by the title will often turn next to the Abstract for a capsule summary of the paper's content. Like titles, Abstracts are widely available and searchable online—so even if your paper is published in a subscriber-access journal, your Abstract will effectively be open-access.

The Abstract summarizes the entire paper, including the research question and its importance, methods, results, and conclusions (and therefore usually adopts a mini-IMRaD structure, although usually without section heads). This summary must be very brief: Abstracts are often limited to two hundred words or even fewer, and rarely exceed five hundred. Therefore, it cannot contain many technical details or much nuanced discussion. The top priority for inclusion is a clear statement of the research question, the (general) approach taken to answering it, and its answer (the main result). Many writers are reluctant to include their results, ending their Abstracts instead with something as uselessly vague as "Results of the experiments are discussed." Perhaps these writers are trying to avoid disrupting the reader's feeling of suspense, but an Abstract is not a movie trailer and does not need to avoid plot spoilers. Readers look to the scientific literature for efficient access to information, and the more they find in the Abstract, the better.

Chapter Summary

- A paper's title is an advertisement, and should indicate the paper's story.
- The way you list your name in your papers' bylines should be as consistent as possible throughout your career.
- The Abstract is a short summary of the paper, including question, importance, methods, major results, and conclusions.

Exercises

1. Examine the Table of Contents for a recent issue of a journal in your field. Which three papers' titles do you find most effective? Why? Which three do you find least effective? Choose one, read the paper, and write an improved title.

2. Choose another paper from the same issue. Don't read the Abstract, but read the rest of the paper. Now write an Abstract. How does your version differ from the actual Abstract? Which is better, and why?

TEN

The Introduction Section

The Introduction as a standalone section of a scientific paper became universal only as the IMRaD structure became the standard in the 1950s, but the function of introductory material has been well understood for millennia. Cicero, writing in 55 BC, identified the purpose of an Introduction as to "attract the hearer straight away" and to provide "either a statement of the whole of the matter that is to be put forward, or an approach to the case and a preparation of the ground" (Volume I, 441). That is, the Introduction combines three functions: advertising, summarizing, and context-setting. Of these, the advertising and summarizing functions have been reduced in importance but not entirely displaced by the addition of the Abstract to the canon. Setting context for the work is now the most important work of the Introduction.

Swales (1990) divides the typical Introduction into three components (numbered in Figure 10.1), which in order move toward the narrow waist of the paper's hourglass:

- **Component 1: Define a research territory**. Here you begin with the hourglass at its widest: a few sentences laying out the broadest possible context for the work undertaken, to establish its importance to a large set of potential readers. For instance, our star-formation paper (chapter 7) might begin with a sentence or two about how life in the universe depends on the existence of stars and planets and thus on the physics by which diffuse gas and dust condense to form such objects.

  The degree of breadth to aim for depends on the target journal. A *Nature* or *Science* paper needs very broad context to justify its appearance in a publication whose scope covers all of the sciences, whereas a paper in a highly specialized journal needs to appeal only

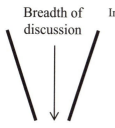

Breadth of discussion

Introduction
General context of the work *(1a)*
Narrower research area and statement of its importance *(1b)*
Identification of a gap or other need for research *(2a)*
Specific research question meeting the identified need *(2b)*
Summary of approach to answer the research question *(3a)*
Announcement of principal findings *(3b)*

Figure 10.1 Structure and components of a typical Introduction. Italicized numbers refer to Swales' (1990) three components: (1) defining a research territory; (2) establishing a niche within that territory; and (3) occupying the niche. The Introduction is shown in the context of the whole paper's structure in Figure 8.1.

to the subset of scientists who read that journal. As an example, consider the first sentences of two recent particle-physics papers:

- From a *Nature* paper on the decay of the strange *B* meson (CMS Collaboration and LHCb Collaboration 2015): "The standard model of particle physics describes the fundamental particles and their interactions via the strong, electromagnetic and weak forces."
- From a *Journal of Physics G: Nuclear and Particle Physics* paper on the detection of fragments from proton-proton collisions (Aaij et al. 2014): "Exclusive J/ψ and $\psi(2S)$ meson production in hadron collisions are diffractive processes that can be calculated in perturbative quantum chromodynamics (QCD)."

The *Nature* introduction starts with a definition of the "standard model" of particle physics, something that situates the paper well for a reader like me, but would be unnecessary (and painfully obvious) for any reader of the *Journal of Physics G*. The *Journal of Physics G* introduction, by contrast, jumps right into the predictions of quantum chromodynamics for meson production in proton-proton collisions. This is fine for the intended readers of that specialized journal, but wouldn't capture many scientists flipping through an issue of *Nature* in the lunchroom.

While not every paper has to matter to everyone, even the *Journal of Physics G* paper sets its work in context beyond the immediate topic: it explains why technical details of particle detection matter to testing an important theoretical model in particle physics. That is, its authors don't just describe what they did; they explain how what they did was important.

Having established context, your Introduction begins to narrow the hourglass by identifying your more specific research territory, along with its relationship and importance to the broader field. For example, having begun your astronomy paper with the importance of star formation in general, you might now indicate that your focus is the formation of massive stars. You might point out that massive stars are critical because their supernova explosions seed newer stars and planets with heavy elements. This part of the Introduction normally includes some literature review to establish the state of the art in the research territory.

- **Component 2: Establish a niche within the research territory**. As the hourglass narrows further, your Introduction works toward your central research question by identifying a research niche: a concrete and narrow open problem within the research territory. This might mean pointing out a gap in our knowledge of some topic. It might mean noting an apparent contradiction in the literature or a published claim that's vulnerable to new information. It might mean identifying rival theoretical models that can be distinguished by new data. It might even mean suggesting an entirely novel way of thinking about a research area. This is where our astronomy paper could point out that there are multiple models for massive-star formation, making different predictions about the appearance and spatial distribution of massive protostars. Finally, your Introduction will state, clearly and specifically, your central research question (whether massive protostars always appear within local clusters of protostars, as the cluster-assist model predicts).

- **Component 3: Occupy the niche**. The third component of the Introduction indicates to a reader how your work occupies the niche you've just identified. In doing so, it further narrows the hourglass by outlining the approach you took to answer the central question, and showing how the answer to this question helps to solve the open problem you identified.

Outlining the approach doesn't mean presenting detailed methods, of course. It means indicating your basic approach (the kind of experiments, observations, and/or theory you executed), the general form of your data (what important quantities you measured, and roughly how), and how analyses of those data can answer your central research question. For our astronomy example, you might indicate that you used radiotelescopic observations to identify massive protostars by inferring mass from the relationship between radius and rotational velocity, and that you calculated distances among protostars based on annual parallax. You would then indicate that this provides a test of the cluster-assist model, because that model is rejected if massive protostars often occur alone.

While my suggestions about the components of an Introduction should so far be uncontroversial, there is considerable disagreement over how best to end one. Some writing guides recommend ending the Introduction with the statement of your central question and approach (e.g., Davis 2005, Katz 2006). Others suggest continuing with a brief summary of your results (e.g., Montgomery 2003, Day and Gastel 2006). My own advice is to include the main result, because doing so shows your reader where you are going and helps signpost their progress through your paper. Don't worry about giving away the ending to your story: you're writing a scientific paper, not a mystery novel, and in any case your Abstract has already ruined the suspense. If you are able to state an interesting and important conclusion, readers closest to your research area will read on to find out how you support it. If other readers are satisfied with a quick answer to your central question, so much the better for all.

Chapter Summary

- The Introduction serves to define a research territory (context), to establish a niche within that territory (knowledge gap), and to occupy the niche (outlining your approach to filling the knowledge gap).
- Establishing context means a broad focus; exactly how broad depends on the target journal.
- A brief statement of your major result is a strong way to end an Introduction.

Exercises

1. Choose a recently published paper in your field, and read the Introduction. Highlight text comprising each of the three main Introduction components (establishing territory, establishing niche, occupying niche).
2. How would you change that paper's Introduction for publication in a journal of narrower scope? One of wider scope?

ELEVEN

〜〜〜

The Methods Section

Methods sections are relatively straightforward to write. They outline the materials you used and the procedures you followed in executing your study and analyzing your data. Often, you can model substantial portions after an earlier paper in which you reported similar work or adapt text from a grant proposal. And most, if not all, of the Methods can be written while you're planning or conducting the research, when the procedures are fresh in your mind. Despite this section's relative ease, however, three issues can be troublesome: organization, level of detail, and avoiding self-plagiarism.

Organizing the Methods

It's tempting to organize a Methods section chronologically, recording what you did in the order that you did it. This might be the right way to write a travelogue or an autobiography, but it's the wrong way to write a scientific paper! Your experience in doing the research doesn't matter; what does is your reader's need to understand it.

There is no one-size-fits-all recommendation for organizing Methods to make them easily understood. One possibility is a three-part presentation of background, experiments or observations, and analysis. This organization begins with material that sets the stage for the main procedures, such as descriptions of your field sites, materials and equipment, methods for selecting subjects, or calibrations or control procedures intended to demonstrate that your procedures work as expected. The second subsection describes your experiments and the data you collected.

The final subsection outlines how you analyzed the data, and how those analyses can answer your research question. This might include such things as the quantities you calculated from the raw data, comparisons you made, or relationships you sought among variables, and the statistical procedures you used to assess any patterns you found.

The background-experiments-analysis organization, though, becomes cumbersome in more complicated papers. If answering your research question involves combining several different lines of investigation (perhaps theoretical and experimental work, or several distinct sets of observations), it's usually better to work though each procedurally distinct component of the work separately. For instance, the story-summary outline for our star-formation paper (chapter 7) included a three-component summary of the observational Methods. With a fourth component added to integrate data from the first three, we have:

- *Determining protostar rotational velocities*
- *Determining protostar masses*
- *Determining nearest-neighbor distances*
- *Testing whether massive protostars always have close neighbors*

An effective Methods section would follow this outline, with four subsections (likely following a few sentences summarizing the overall approach). Each subsection could separately follow the background-observations-analysis organization above: the first, for instance, would specify the nebulae observed, describe the instruments used to measure red/blue shifts of the ^{13}C emission line, explain how those shifts were measured, and outline how rotational velocities were calculated from the shifts. This organization works because the four subsections are logically separate, but each builds on the ones before. (Mass determination, for instance, relies on the results of the rotational velocity measurements.) An alternative arrangement introducing instrumentation for all components, then describing all the observations, and finally dealing with data analysis would force the reader to shift attention repeatedly from velocities to masses to distances and back again.

Whatever organization works best for your Methods, it's a good idea to signal it clearly with a system of subheads. (The four bullets above would work well for our star-formation paper.) Subheads enhance your paper's finding system, orienting the reader with respect to the larger argument and identifying components that can be tackled and digested relatively independently. It's especially effective to match subheads in the Methods with identical subheads in the Results, so that the reader can easily navigate back and forth.

Appropriate Detail

In writing your Methods, you must decide how much detail to supply for each procedure. Getting this right can be tricky. Some details must obviously be included: for instance, journals nearly always require mention of ethics-board approvals for work involving human subjects. Other details should obviously be omitted: whether you took your notes with a 2H or HB pencil is utterly unimportant (although, believe it or not, I've seen manuscripts specifying this). Between these extremes is a long continuum of relevance and plenty of grey area.

How do you decide whether to include or omit a given detail? Experts offer different answers to this question because they disagree about the function, for the reader, of the Methods section—a function that has evolved considerably over the last 350 years (Box 11.1). Most writing books (e.g., Katz 2006, Day and Gastel 2006) tell you to give readers enough detail so that they could repeat your work and verify the results themselves. However, studies of the way scientists *actually* write find that few published papers come close to this level of detail (e.g., Swales 1990, Gross et al. 2002). (The main exception to this generalization is the "methods" paper, in which providing new methodology for others to repeat is the major point.) These studies suggest that your Methods section is best seen as establishing the credibility of your approach, and thus giving readers a reason to believe your findings. In other words, "if critical readers judge . . . [the Methods] a plausible strategy for solving the problem stated in the Introduction, then they will likely view the article as authentic science" (Harmon and Gross 2007, 193). In addition, the Methods tell readers what they need to know about the procedures if they are to understand the Results.

Box 11.1 Replication, witnessing and authority: the evolution of the Methods

The past 350 years have seen considerable evolution in the way methods are communicated, which reflects underlying evolution in the *reason* for relating those methods. This evolution accounts for much of the modern disagreement over the level of detail Methods sections should include.

Scientists working at the birth of modern scientific communication, in the seventeenth century, belonged to the intellectual tradition of the European Renaissance. They believed learning should come from empirical observation rather than from study of earlier texts (as in the medieval "scholastic" tradition). But as science progressed, it became more and more obviously impossible to make progress without building on results reported by others. So how could those reports earn authority?

This question was a major concern for Robert Boyle, the pioneer of the modern scientific paper in the 1660s (chapter 1). Boyle's answer was threefold (Shapin 1984). First, he favored exhaustive detail of equipment and procedures, so that readers could repeat his experiments for themselves. Second, Boyle argued for "communal witnessing": if one was to rely on the results of others, then those results must be witnessed by other scientists. Thus, many of Boyle's key experiments were conducted in public, and Boyle published the names and qualifications of his witnesses along with his results. Third, Boyle described in exhaustive detail not just his methods, but his experiments' circumstances and settings, his false starts and failures, and much else. Illustrations of his experimental apparatus were detailed and realistic depictions, not simplified line drawings. For example, to accompany his reports of experiments using his famous vacuum pump (Boyle 1660), he provided an illustration of the *particular* pump he had used, complete with irregularities,

dents, and dings. The point of all this description was to make readers feel as if they had been there—to recruit them as "virtual witnesses" (Shapin 1984).

This approach was widely adopted. A spectacular example is Pierre-Louis de Maupertuis' (1737) account of an Arctic expedition to measure the Earth's shape. Maupertuis spends many pages relating the excitement and hardships of his travels. He describes, among other things, the midnight sun, the assaults of biting flies, techniques for defense against kicking reindeer, and cold that left only his brandy unfrozen to drink.

None of Boyle's three answers to the authority problem proved fully satisfactory. Few others ever tried to repeat his experiments. Communal witnessing was cumbersome and inefficient. Virtual witnessing had more to do with rhetoric than logic, and it made publications verbose and unwieldy. It was gradually abandoned.

By the mid-nineteenth century, the professionalization of science led to a new kind of authority. A report began to be considered reliable not because it was repeatable, witnessed, or detailed, but instead because its author belonged to a community of established scientists and had professional credentials and/or an institutional affiliation. Along with this came an emphasis on detachment and objectivity in writing, which meant depersonalizing and simplifying the Methods (Daston and Gallison 2007). Authority also came from peers recognizing a scientist's use of standard or appropriate methods. In the early twentieth century, peer review became standard; it gave reports further authority because they had passed muster with experts who consider appropriateness of method, but almost never attempt to repeat it.

In modern science, repeatability and witnessing both survive, but their role lies largely in testing extraordinary claims such as cold fusion (Fleischmann and Pons 1989) or hyperdilution memory of water (Davenas et al. 1988). Professionalism is now the major grounds for the authority of most published work, which explains why cases of scientific fraud are always shocking and often slow to be discovered. (For instance, the psychologist Diederick Stapel falsified data for at least fifty-five publications before his misconduct was identified [Stapel Committee 2012]). Fortunately, fraud appears to be relatively uncommon—estimation is difficult, but between 0.001 percent and five percent of scientists appear to have falsified data at least once (Fanelli 2009)—and seldom distorts accepted understanding for long. This suggests that the lack of emphasis on repeatability is not a major handicap to the progress of science.

Thinking of the Methods as establishing credibility suggests that a detail merits inclusion if it fulfills one of three slightly different functions. First, it might establish your qualification as a researcher (that you know how to use standard methods in appropriate ways). Second, it might establish the plausibility of your approach to the problem (that you are gathering relevant data and analyzing it in a way that sheds light on your research question). Third, it helps establish your sequence of investigative steps, so that the reader can understand the logical basis for claims to come in your Results and Discussion. A simpler expression of this might be that a detail should be included if, and only if, it could influence the reader's interpretation of your Results.

Take a real example[1]. A student of mine, Chris Kolaczan, studied genetic variation in a parasitic wasp, one that attacks a caterpillar living inside a gall on the stem of a goldenrod. Chris's published Methods section (Kolaczan et al. 2009) covers field collection and preservation of the wasps, DNA amplification, fragmentation, fragment-length determination in the laboratory, and analysis of the resulting data. It includes the following sentence:

> At each collection site, we collected and opened . . . galls, removing the [caterpillars] and inspecting them for the presence of . . . [parasites]. Parasitized [caterpillars] were immediately preserved in 95% ethanol.

Here, Chris describes opening galls and identifying parasitized larvae. This contributes to the plausibility of the approach and gives the reader enough knowledge of the series of investigative steps to understand Chris's results. For instance, it matters that each wasp analysed came from a different caterpillar in a different gall. For someone looking to repeat the work, though, a lot is missing. The collection sites are located (by an Appendix) only to within about 1.5 km, and Chris doesn't specify how to recognize a galled plant, open a gall, or distinguish a parasitized caterpillar from an unparasitized one. Chris also mentions that preservation was immediate and done with 95% ethanol, establishing that he used standard protocols to adequately preserve tissue for DNA analysis. That is, Chris attests that the reader needn't worry about

[1] While I've been enjoying my star-formation example, engaging too closely in the details of that hypothetical study risks revealing that I don't actually know very much about the topic. We wouldn't want that.

some obvious problems that might compromise interpretation of the results (for instance, artifacts arising from degradation of DNA before analysis). These preservation details are not necessary for someone to repeat the work (because alternative methods, such as freezing in liquid nitrogen, would be equally suitable), but they contribute to credibility. Chris *doesn't* mention the containers in which preserved larvae were held (4 mL polypropylene vials, actually), because it's hard to imagine a reader's reaction to the study depending on whether larvae were kept in plastic vials or glass ones (or coffee cups, for that matter). Overall, this Methods section did not provide enough detail for someone to repeat the work exactly, but it did establish Chris's authority as a scientist, the plausibility of the approach, and the sequence of investigative steps. As a result, readers should find the results credible and understandable.

The credibility criterion for inclusion of detail is less cut-and-dried than the repeatability criterion, but it can be applied with some careful thought and familiarity with the literature. Reading published papers in your field is enormously helpful, because each field (and sometimes each journal) has conventions for the level of detail expected in Methods. For example, in ecology it's standard practice to identify the software package used for statistical analyses, but in cell biology, this detail is rarely reported. While these conventions can seem arbitrary, noticing and following them helps you meet reader expectations and build credibility.

What of a reader who actually does want to repeat your experiment? Such readers are important, but very rare: a vanishingly small fraction of published studies are ever exactly replicated by another scientist (Casadevall and Fang 2010, Loscalzo 2012)[2]. Let us imagine, extremely generously, that one percent of your readers want to repeat your work using your methods; the rest want only to be assured of the credibility of your results. Providing the wealth of detail desired by the one percent would reduce clarity and increase reading effort for the vast majority. And it wouldn't make repetition more likely anyway: its infrequency has little

---

[2] If scientific results aren't routinely verified by repetition, how are they verified? Many never are, and in this collective shrug science indicates its opinion of their importance. The rest are verified because their results prove consistent with other results, and because other scientists are able to build further understanding on top of them.

to do with writing style, and much to do with the time and expense required to replicate work and the lack of professional rewards for doing so. Very detailed procedures can be provided in an online supplement (chapter 14), or can simply be omitted, because would-be repeaters can contact an author for further detail.

Avoiding Self-Plagiarism

It's common to find yourself writing multiple papers that use the same methods or study system. Most scientists easily recognize that reusing the same data or analyses in multiple publications is inappropriate, but what about reusing descriptions of one's Methods? Writers are sometimes surprised to learn you can model new text after old, but that simply reusing old Methods text is a form of plagiarism. If you're skeptical about this, consider Wei et al. (2010), which was retracted by the journal, over the authors' objections, after a reader noticed substantial repetition of Methods (and Results) from an earlier paper (see Retraction Watch, http://bit.ly/1SaJBTy).

Plagiarizing yourself is something of a tricky concept. You can't steal your own silverware, so how can you steal your own words? Well, you may own your silverware, but you usually don't own your published words—instead, copyright tends to be assigned to the journal's publisher. Legally, therefore, you are no more free to reuse your own words without permission than you are to reuse mine. (In contrast, reusing words from a grant proposal is perfectly legitimate, as proposals are not considered published.)

When you need to describe a procedure again, and can't repeat a previous description, you can use one of two techniques. Sometimes a later paper can include only the bare bones of a method, citing your older work for more detail. This might be the case, for instance, when the first paper describes a novel technique, but the second need only assure its readers that a technique is available. More often, though, you'll want to repeat some detail because you can't expect readers to dig out your previous papers to understand your latest one. Then you will have to rephrase your earlier Methods. Fortunately, English is a rich enough language that there is never just one good way to express something. As an example, these short passages come from the Methods sections of two papers that needed to introduce their readers to the same study system (a pair of goldenrod species and the insects that feed on them):

> The goldenrods *Solidago altissima* and *S. gigantea* are clonal perennials codistributed over much of eastern and central North America. Intermixed stands of the two species are common in open habitats such as prairies, old fields, roadsides, and forest edges. Individual ramets grow in spring from underground rhizomes, flower in late summer and fall, and die back to ground level before winter. . . . *S. altissima* and *S. gigantea* are attacked by a diverse fauna of insect herbivores, which vary in diet specialization (Heard 2012; citations and a few details removed for clarity).

> The goldenrods *Solidago altissima* and *S. gigantea*, two closely related members of the *Solidago canadensis* complex, share a diverse herbivore fauna. *S. altissima* and *S. gigantea* are abundant and frequently syntopic in prairies, old fields and disturbed habitats across much of temperate North America. They are long-lived rhizomatous perennials, with new ramets growing from overwintered rhizomes each spring, flowering in late summer to fall, and senescing to ground level in late fall (Heard and Kitts 2012).

Comparing the two passages, notice three things that helped us avoid self-plagiarism. First, each includes a slightly different set of details about the system. This is partly because different details were important to the two papers (variation among herbivores in diet specialization was central to the first paper, but not to the second); but it also helps keep the writing fresh. Second, the information is ordered differently, something that would be even more apparent from longer passages. Third, even where the same information is presented, the words and phrasing are fairly different.

Chapter Summary

- Methods have many possible organizational schemes, but a chronological narrative is rarely effective.
- Sources disagree on the detail necessary in Methods because they disagree on the section's function.

- Most readers will not try to reproduce your work; therefore, it is not necessary (or wise) to provide the level of detail necessary for someone to repeat your experiments exactly.
- A detail should be included in Methods if it (a) establishes your qualifications as a researcher; (b) establishes plausibility of your approach to the problem; or (c) helps establish the investigative steps so a reader can understand your solution of the problem.
- You may have to rewrite Methods text from paper to paper, because you can't repeat your own published wording without copyright issues.

Exercises

1. Take an experiment (or observation) that you've planned or recently completed, and write two versions of Methods text to describe it. In the first version, include enough detail that another scientist could reproduce your work exactly. In the second, include only enough for readers to understand your work and find it credible. Which version seems closer to the Methods sections you see in the literature for your field?

TWELVE

The Results Section

Writing the Results seems as though it should be easy. Its contents certainly seem obvious: this is where you report the results of your experiments (or observations, or theoretical work). True enough; but this isn't a very useful way of thinking, because it emphasizes the writer's experience, not the reader's. Writers who take this perspective produce manuscripts bulging with preliminary results, multipage tables, dozens of complicated graphs, and every data point from every experiment they ever ran. The rare reader who sticks it out is left torn between bewilderment and resentment—and this isn't where you want *your* reader to be. So remember the importance of finding and telling your story (chapter 7), and strip your Results down. Show only the data the reader needs to understand and accept the answer you're presenting to your central research question. What remains will be a short and simple section in which every word, graphic, and data point contributes directly and obviously to telling your paper's story.

With content set, what remains is presentation. The Results section must tread carefully in its relationship to the Methods that precede it and the Discussion that follows. It must be organized so its most important content is easily apparent to the reader. Finally, it needs to communicate complex and heavily quantitative information while remaining easy to read.

Methods, Results, and Discussion

The strict separation of Methods, Result, and Discussion can seem forced. It might seem easier, and more natural, to write "I did this, and this was the result, and this is what it implies. Then I did that, and got these data, which I interpret this way. Finally I did this, with this result, which means the following." Actually, this *is* easier for you as the writer—but for two reasons it makes things harder for your reader (which is the more important consideration). First, relatively few readers will read Methods, Results, and Discussion with equal interest. Some skim or skip the Methods, some read Results but ignore Discussion, and some read Discussion but have little patience for detailed Results. Others may want to access one section directly: the Methods to borrow techniques, or the Results to add your data to a meta-analysis. Second, even if integrating all the sections worked well in principle, the canonical organization would remain superior simply because it *is* the canon: your readers are accustomed to this organization. When you use it, you meet their expectations and put information right where they will look for it.

Separating Results from Discussion does not mean presenting data entirely without comment. A good Results section draws your reader's attention to features of the data that you will later interpret. If you write "average dry mass was 14.2 ± 1.1 g for fertilized plants and 9.4 ± 2.3 g for unfertilized plants," you ask your reader to figure out what pattern there might be. Writing "fertilized plants grew over 50% faster than unfertilized ones (average dry mass 14.2 ± 1.1 g vs. 9.4 ± 2.3 g)" highlights a contrast that readers can relate to your story. The Results section can draw comparisons between controls and treatments, between experiments, or among experiments, observations, and theory. *Interpretation* of those comparisons, and comparisons with literature results, should generally be kept for the Discussion.

Despite my argument for separating Results and Discussion, the most common deviation from the IMRaD canon is their merger into a single section. Some journals allow this, others don't, and a few (notably *Nature* and *Science*) encourage it. If your target journal allows a choice, ask yourself the by-now familiar question: which arrangement favors the clearest communication with your reader? Sometimes, in a long paper with many sets of results, a merger reduces repetition and lets you discuss results that are fresh in the reader's mind. If you are tempted by such a merger, though, be sure of two things: first, that the length of your paper doesn't betray a failure to find your story; and second, that the advantages of integrating Results and Discussion outweigh the cost of compromising your reader's finding system.

The separation between Methods and Results is much closer to a fire-wall. Your Results section may *mention* methods, but it should never *introduce* them. A reminder of where results came from can be useful: for instance, "Plants fertilized with slow-release micronutrients grew 50% better than unfertilized plants and 22% better than plants fertilized with micronutrients in aqueous solution." But such reminders should be brief and used only when a reminder of the methods is important to the reader's understanding of your results. In the other direction, you might occasionally mention a result in a Methods section, but only when it's needed for your reader to understand or accept the methods used, and it isn't discussed further. For instance, your Methods might read, "We used parametric ANOVA to test for differences between treatments. This approach is justified because our data showed no significant deviation from normality of residuals or homoscedasticity."

Carefully separating your results from your methods and discussion may leave a Results section that is startlingly short—perhaps a few paragraphs or even a few sentences (plus, of course, tables and figures). This does not mean that your work is trivial or your writing simplistic: it means that you have found an elegant approach to a well-defined research question, and presented it clearly to your grateful reader.

Organizing the Results

Even a very short Results section can benefit from careful organization, and longer ones require it. When your data and analyses are fairly simple, you do well to place the main result (the one that most directly answers your central research question) in the first paragraph. This is a "power position," in the sense that readers tend to emphasize material found there. Subsequent paragraphs can include data and analyses that support or complement your main result.

This main-result-first organization isn't feasible, however, for papers that make more complex arguments. Often, your "main" result is really a synthesis of several lines of evidence, or involves building later analyses on the results of earlier ones. In our star-formation paper, for instance, the main result (the fraction of massive protostars having close neighbors) depends on two intermediate results (protostar masses and nearest-neighbor distances), and the protostar masses in turn depend on rotational velocity measurements. Here it is more effective to work through the results from least to most complex, following the structure of your Methods section and using identical subheads. This organization brings the reader along as you build logically to your main result at the end (another power position). Your reader will appreciate a signal that they've reached that main result—something like "Finally . . . ," "Most important . . . ," or "Combining the results so far, we arrive at a test of our main hypothesis. . . ."

Perhaps the strongest organization for complex Results hybridizes the main-first and main-last techniques. Here, you work from basic to complex, but precede that with a very brief overview of the most important result. For our massive-star formation paper, we might open the Results with "Our analyses, taken together, show that only a moderate fraction of massive protostars occur in tight clusters . . ." before backing up to present the separate results that combine to establish that conclusion. The advantage of tipping one's hand at the beginning of the Results is that the reader knows in advance the full argument into which each result will fit.

Communicating Quantitative Information in Text, Tables, and Figures

Results sections are, almost without exception, numbers-dense, but without skillful handling, numbers make text dense and difficult to read. Tables and figures (I'll refer to these collectively as "graphics"), when handled well, allow efficient and reader-friendly display of numbers, and therefore will be important to almost every Results section you write. Graphics need not be quantitative and can appear anywhere in a paper; but the vast majority serve to display numbers in the Results, so I deal with them here.

Not every number belongs in a graphic, and whether placement in text, table, or figure works best depends on what it is you need to show your reader. When you present just two or three numbers to make a point, place them directly in the text, where they won't interrupt your manuscript's flow. Using a graphic imposes a cost of navigation on your reader, who must divert attention away from the text, find and inspect the graphic, and then find their place in the text again to pick up the thread of your argument. For just a few numbers, placement in text is also much more compact than a graphic requiring legend, labels, and white space within and surrounding it. However, if you need to present more than a few numbers, placing them in your text can make your prose indigestible and the pattern obscure. Here, a graphic's ease of reading will compensate for the navigation cost of accessing it. This tradeoff is illustrated in Box 12.1.

Box 12.1 Numbers in text vs. graphics

For displaying two or three numbers, a graphic wastes page space, and asks your reader to divert attention from text to graphic and back again: *"Fertilized seeds were larger than unfertilized ones (Figure A)"*. This could have been presented more effectively in text as *"Fertilized seeds were larger than unfertilized ones (1.8 ± 0.3 g vs 0.9 ± 0.2 g)."*

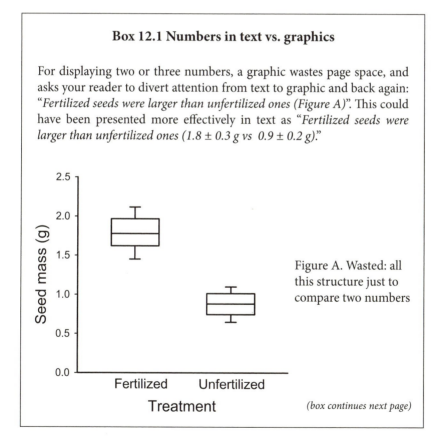

Figure A. Wasted: all this structure just to compare two numbers

(box continues next page)

(cont.)

Graphics pay dividends, though, for larger sets of numbers, because the help they give in seeing pattern outweighs the cost of diverted attention. Imagine being asked to wade through this: *"For lakebottom sediments collected at the inflow, average grain size was 1.3 mm (± 0.4 mm standard deviation). Smaller grains predominated farther out: average 0.8 mm (± 0.2 mm) at 5 m from the inflow, 0.2 mm (± 0.05 mm) at 10 m, 0.08 mm (± 0.03 mm) at 20 m, 0.012 mm (± 0.004 mm) at 40 m, and 0.002 mm (± 0.0005 mm) at 60 m."* Ugh! The pattern is easily seen in a figure: *"Grain size of lakebottom sediments decreased away from the inflow (Figure B)."*

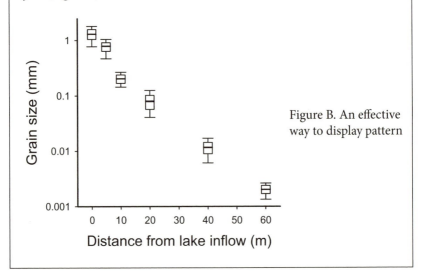

Figure B. An effective way to display pattern

Among graphics, tables excel at presenting datasets with many entries and/or variables (presuming that your story needs them). That's because adding more rows or columns to a table detracts little from the readability of what's already there. Tables are also the method of choice when your reader needs precise values of numbers. The major disadvantage of tables is that they aren't an effective way to show trends in data or relationships between variables (Box 12.2). Therefore, reserve tables for sets of numbers that stand individually (for example, a list of a mineral's properties—composition, hardness, cleavage, and so on), or for datasets where each reader will want to look up a few particular numbers (e.g., a list of molecular weights).

Box 12.2 The main weakness of the table

It's difficult for readers to see patterns or relationships among variables when you present data in a table. This dreadful example may be the worst sin I've ever committed as a writer (Heard and Remer 1997, but the blame is all mine):

TABLE 2

COEXISTENCE TIME AND CLUTCH SIZE FOR EQUAL COMPETITORS

CLUTCH SIZE, SPECIES 1	CLUTCH SIZE, SPECIES 2				
	1	2	4	8	16
1	**134** (14)	82.3 (4.6)	49.5 (2.2)	27 3 (1.3)	16.9 (.8)
2	80.6 (5.5)	**145** (13)	60.2 (2.6)	31 5 (1.4)	17.4 (.7)
4	48.4 (2.5)	59.6 (3.0)	**182** (16)	41 0 (1.9)	19.4 (.9)
8	28.5 (1.5)	31.5 (1.5)	43.3 (2.4)	**520** (53)	26.6 (1.2)
16	16.6 (.6)	17.4 (.7)	20.3 (1.0)	26 0 (1.4)	**2,442** (213)

Note. — All parameters as in table 1, except $\alpha_{12} = \alpha_{21} = 1$. Coexistence times are as follows: bold, either species excluded at random; underlined, species 1 excludes species 2; regular font, species 2 excludes species 1. Numbers in parentheses are twice the standard errors.

I wanted to show that "coexistence times" are longer for equal (main diagonal) and larger (lower right) "clutch sizes." And I suppose you could figure that out, if you worked really hard at it—but who would? I used too many numbers with too much precision, and the underlining and boldfacing didn't help nearly as much as I thought it would. A figure would have shown the trends much more clearly.

Figures excel at communicating trends in data and, especially, relationships among variables. This is particularly true for relationships that are nonlinear: these are nearly impossible for readers to discern from a table, but readily apparent in a figure. However, figures do not display precise numbers well. They also do poorly at displaying more than a few variables at a time: adding more plots to your graph can quickly produce an indecipherable tangle of lines or forest of bars.

In summary, if you have just a few numbers to present, embed them in your text; for more, use a graphic. Among graphics, use a table to convey exact values, and a figure to illustrate trends and relationships. For any set of numbers, though, choose just one option: repeating the same information in text, table, and figure wastes your reader's attention and your publisher's resources.

Handling Numbers

Because numbers ask a lot of your reader, you should work hard to minimize those demands. Here are some ways to do that.

- **Winnow data for presentation.** Your job as a writer is not to blow the reader away with a numerical tornado; it's to present just enough data to tell your story convincingly. If several variables are strongly intercorrelated, display the most relevant one, or use a data-reduction tool such as principal components analysis. If there are several metrics that could quantify something, or several statistical analyses that could test for a pattern, present the one most applicable to your research question. If readers might question whether your results are robust to your winnowing, you can place alternative variables, analyses, or metrics in online supplements (chapter 14) or simply say, "Use of alternative metrics, such as this-and-that, yielded similar results."
- **Omit redundant numbers.** Don't inflate number content by providing several versions of what is fundamentally the same number. For example, suppose that I wrote "Among massive protostars, 13 of 25 (52%) had close neighbors, while for small protostars just 6 of 25 (24%) did." Here counts and percentages are redundant, because the denominators are the same so either can be compared easily. Along similar lines, writers reporting analyses of variance often provide degrees of freedom, sums of squares, and mean squares. But any of these three is easily calculated from the other two (MS = SS/df), so presenting two always suffices.
- **Emphasize the most important numbers.** When you do report multiple sets of numbers for a single result, let your reader know which are key and which play a supporting role. For example, imagine that you're comparing two fractions with different denominators, so it's helpful to provide percentages. There are two ways to phrase this: "25/46 vs. 23/67 (54% vs. 34%)," or "54% vs. 34% (25/46 vs. 23/67)." Use the former if you want the reader to emphasize counts and sample sizes, but use the second if you want the reader to compare percentages.

A common emphasis problem arises in presenting statistical results. For instance, we could compare the fractions from the last paragraph, and write either "A χ^2-test showed that massive protostars were more likely to have neighbors, with $\chi^2_{(1)} = 4.47$, $P = 0.03$ (54% vs. 34%)" or "Massive protostars were more likely to have neighbors (54% vs. 34%; $\chi^2_{(1)} = 4.47$, $P = 0.03$)." The first wording emphasizes numbers associated with the statistical test, which play a supporting role but aren't what your reader picked up your paper to see. The second and much better wording emphasizes comparisons that tell your story.

- **Report only meaningful and necessary precision.** When you report a number, how much precision—"1.68234119478," "1.6823," "1.7," or "about 2"—should you provide? The answer depends in part on the number's "significant digits" and in part on what use the reader will make of it.

    Significant digits are those digits in a number that you know with reasonable confidence. These are determined by (1) the precision of the measurements that yielded the number; and (2) the propagation of uncertainty during calculation (when a number is derived from multiple measurements, each with its own uncertainty). If air currents in your lab make your balance precise only to the nearest 0.1 g, it would be silly to report that a seed had a mass of 1.6823 g; make it 1.7 g. Similarly, don't report a mean seed mass of 1.6823 g if the standard error is 0.01 g; make it 1.68 g. There are technical guides to determining significant digits (e.g., Robinson et al. 2005, Ch. 1; or http://www.hccfl.edu/media/181113/sigfigs.pdf), but as a rule of thumb, a digit is significant if you'd expect it to be consistent through multiple measurements or calculations of the same quantity. Reporting more digits than are significant makes your manuscript more difficult to read without adding any compensating information.

    Significant digits set the *maximum* precision you should report. However, the reader may not need that much precision to understand the story you are telling. If you can measure seed mass to the nearest 0.0001 g, should you report seed masses of 1.6823 vs. 0.7714 g for fertilized and unfertilized plants? Would this tell your reader more than 1.7 vs. 0.8 g? In most situations, it would not, and the rounded numbers are better.

    Statistics seem to cause many writers particular trouble with precision, probably because software tends to report test statistics and P values to many (unwarranted) decimal places. I frequently see manuscripts with statements like "seed mass did not differ significantly among treatments ($F = 0.92238674$, $P = 0.7826$)." Nearly all those digits mean nothing to the reader! "$F = 0.9$, $P = 0.8$" carries just as much information and does so more clearly. Two significant digits are normally plenty for P values, and two or three for most test statistics.

Designing Tables

There are many choices for arrangement of information in a table's rows and columns. Tables that are not carefully designed can be impenetrable. A full treatment of table design is beyond the scope of this book, but attention to a few basic principles can greatly improve communication with your reader. (For further guidance, see e.g., Tufte 2001, Council of Science Editors 2006.)

- **Arrange tables to take advantage of natural reading patterns**. Because English text is read from left to right, your readers will tend to follow the same pattern when inspecting a table. Therefore, place familiar or context-setting information in the leftmost column, and new or dependent information in columns to the right. For example, place variable names at left and their values to the right; independent variables at left and dependent variables to the right; or pretreatment conditions at left and posttreatment measurements to the right. Use similar logic to order rows from top to bottom.
- **Showcase patterns with vertical rather than horizontal display**. Readers will find it easier to see patterns if they can compare entries you've arranged vertically. Our positional notation system makes this especially true for numbers:

1,359, 11,280, and 104,600 or	*1,359*
	11,280
	104,600

For similar reasons, tall narrow tables are easier to read than wide shallow ones. Keep in mind, though, that if you're asking your reader to see a complex pattern, a figure is probably better.
- **Format tables for easy reading**. Use design tools such as lines, white space, and indenting to draw the reader through your table's organi-

zation. Separate rows and columns well, lest you inflict on your reader an unbroken sea of numbers (or other table entries). Most journals allow lines between rows, but not between columns; use white space instead. Label rows and columns clearly, with a minimum of cryptic abbreviations, and if you want your reader to compare two rows or columns, place them beside each other.

- **Keep tables as few and small as possible.** Columns, rows, and even whole tables are so easy to create that they seem to multiply when you're not looking. However, even well-designed tables demand that your readers navigate to them and locate material within them (the latter becoming more difficult as a table grows). Keep only the essentials. Tables that need to be reproduced in landscape format or that span multiple pages are particularly difficult for readers. Finally, don't try to compress tables by using smaller fonts or removing white space; if your table won't fit on a page, compressing it will only make it unreadable.

Designing Figures

Figures come in a bewildering array of types: maps, photographs, line drawings, scatterplots, boxplots, pie charts, ternary plots, and dozens (if not hundreds) of others. Software makes it easy for you to create any of these, but offers you little help with choosing among them or designing the figure well. The overarching principle of figure design is to give your reader "the greatest number of ideas in the shortest time with the least ink in the smallest space" (Tufte 2001, 51). As with tables, a complete treatment is beyond the scope of this book, but some basic principles are worth emphasizing.

- **Use straightforward and familiar types of figure.** Makers of graphics software trumpet their packages' abilities to depict data in ever-more-dazzling ways. Bars can be grouped, stacked, color-coded, or built from icons; pie charts can be exploded; three-dimensional plots can be kriged, rotated, or heatmap-coded. But the fact that you *can* do something doesn't mean you *should*: glitzy and novel approaches to figure design are more likely to impede communication than facilitate it. Whenever possible, choose figure types that emphasize your data rather than the cleverness of your graphics software; these are usually types that are straightforward in construction and familiar to readers.

Figure types fall into three classes (with some overlap). Data reproductions are minimally processed representations of actual observations (photographs, instrument tracings, etc.). Schematics are simplified or abstracted material such as maps, flow charts, and line drawings. Data compilations display and summarize numerical data (scatter plots, bar charts, and so on).

Data reproductions should be used sparingly in written materials (although they are quite useful in illustrating talks). A photograph or instrument tracing is generally just a single datum, from which (alone) little can be inferred. Furthermore, data reproductions almost always include extraneous detail that captures reader attention but doesn't contribute to telling the story. Data reproductions can be useful as exemplars, giving readers a feel for the study system or the nature of the data. Their moderate use is conventional, furthermore, in some fields: papers in cell biology, for instance, often reproduce gels or stained tissues, and papers in in biosystematics often use photographs to illustrate features of typical specimens.

Schematics are generally used as visual support for in-text explanation. They can appear in any part of a paper: for instance, drawings of apparatus or flow charts summarizing algorithms often appear in the Methods. In the Results, schematics might include things like line drawings depicting features of fossils, or maps or diagrams of geological formations. The great advantage of schematics over data reproductions is their abstraction: they can remove detail to emphasize relevant features or extract generalizations from noisy observations. Of course, this means that they are interpretations rather than data, reflecting decisions by their makers that customize them to the story being told.

The line between data reproductions and schematics can become blurred, as when photographs are enhanced or retouched for publication. Such enhancement, when clearly disclosed, can do readers a great service: for instance, increasing the contrast of a photograph may make relevant features (such as faint stars in a telescopic view or bands on a DNA-fragment gel) easily seen. Undisclosed, the same manipulation would be misleading at best and fraudulent at worst.

Data compilations account for the large majority of the figures we use. They come in many different types, which are best suited to highlighting different kinds of patterns. For instance (Box 12.3), scatterplots with fit lines are good for showing nonlinear relationships between two variables; box plots are good for showing comparisons between average measurements, and divided-bar charts for accurately depicting parts of a whole. Choosing an effective data-compilation type, therefore, means deciding which pattern, comparison, or other feature of the data you want your reader to emphasize. If, having decided this, you are still not sure what type to use, consult

Box 12.3 Some simple data compilations

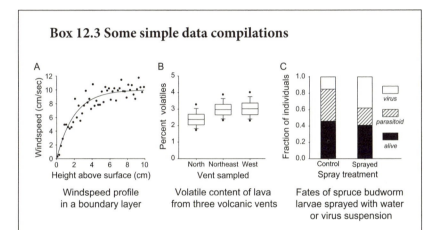

A — Windspeed profile in a boundary layer

B — Volatile content of lava from three volcanic vents

C — Fates of spruce budworm larvae sprayed with water or virus suspension

These three graphs illustrate simple but effective kinds of data compilation. In each case, I've chosen a graph type that does a good job of highlighting the pattern I want the reader to see. Graph A shows wind speeds through the boundary layer above a solid surface; the scatterplot communicates variability while the fit line emphasizes the asymptotic nature of the curve. Graph B compares volatile content (water vapor, CO_2, etc.) of lava from three vents of a volcano; the boxplot makes it easy to see differences in means and spread. Graph C compares the fates of spruce budworm larvae in two virus-exposure treatments; the divided bars make it easy to see that the larvae killed by virus would mostly have been killed by parasitoids anyway. Many more graph types are available, each with its own advantages.

a technical guide on figure-making (Tufte 2001, Kosslyn 2006) or read the literature to find figures that effectively communicate patterns like the one you wish to show.

- **Make figures simple**. Software doesn't just make it easy to make figures—it makes it easy to make them complex, with multiple panels and many data traces (plots for different treatments, sets of observations, or variables). I recently reviewed a manuscript that included seven figures with an astonishing forty-three panels and eighty-four data traces—and for readers craving more, provided another eight figures, fifty-four panels, and sixty-eight data traces in online supplements. Another manuscript included a single figure that had twenty panels and ten different data-point symbols defined in four keys (Box 12.4). Perpetrators of figures like these don't understand that it's the writer's job, not the reader's, to extract pattern from data and relate it to the story being told. As a rule of thumb, a single figure should rarely have more than four panels (fewer is better), and a single panel should rarely have more than four data traces. If your figures are more complex than this, think seriously about breaking them up or deleting some elements (or at least moving them to online supplements where they can be more conveniently ignored).

- **Keep figures readable**. Most figures are reduced in size during publication, usually to half (or less) their sizes in your draft manuscript. To compensate, use thick lines and large fonts and symbols, and check to make sure figures are easily readable after such dramatic reduction. With electronic publication, readers *could* zoom in; but don't require them to do so, as it will disrupt their reading momentum. If using large elements makes your figure cluttered or densely packed, don't shrink the elements; simplify the figure.

    When coding data points, lines, curves, and areas to a key, use symbols that are easily distinguished (such as filled and hollow circles or solid and dashed lines). Avoid subtle distinctions (triangles facing up vs. down, lines differing in thickness). Use consistent conventions across panels and figures: identical fonts, symbols, bar widths, axis scaling, and so on. Every shift in design between figures (say, filled vs. white bars in one figure and hatched vs. white in the next) asks the reader to master a new visual dialect. Provide symbol keys in the fig-

Box 12.4 Abusing your reader with a figure

This figure shows fish species diversity as a function of lake area, for several sets of pollutant-exposed lakes in Quebec, and for historical surveys over one hundred years. Or at least, it attempts to show all that. Its designer asks the reader to inspect patterns in twenty panels—five each for four sets of lakes. It's not obvious whether what's important is the overall picture (most panels show increasing trends), comparisons among panels within sets (slopes mostly decrease through time), or comparisons of corresponding panels among sets. Furthermore, the use of distinct symbols to key data points to their corresponding lakes suggests that perhaps the reader is being asked to think about each lake individually—but without being told why. (While I've disguised the real source of this figure, I haven't exaggerated its complexity.)

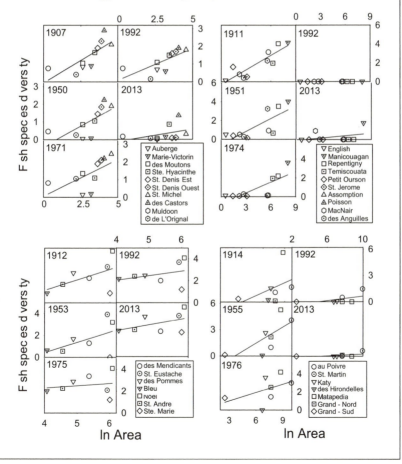

ure itself, not in the legend where they ask the reader to switch attention repeatedly between visual and text elements. Whenever possible, set axis scaling so that data occupy the whole plane of the figure (unless you are reserving white space for a symbol key, or unless white space results from keeping consistent axis scaling among panels or figures).

- **Minimize color**. Until quite recently, color figures were very expensive to produce and, not coincidentally, very rare. The recent trend to online publication has made color figures cheap and easy to publish—but that doesn't mean they're a good idea. While color can be useful in some situations (e.g., heat-maps for three-dimensional information, or showing multiple staining of biological tissues), it has significant disadvantages that are often overlooked. First, about three percent of the population have color-vision deficiencies, with red-green colorblindness the most common. These readers may not perceive color contrasts that you intend to carry information. Second, colors (and color contrasts) will shift depending on the frequency spectrum of ambient light and characteristics of the displaying device. Third, color figures published online are not always read there. For the foreseeable future, you should expect some readers to print your paper (most likely in black and white) before reading it. These issues make color figures advisable only when there is no other way to communicate the necessary information.

- **Don't let figures mislead readers**. Figure design always involves choices that affect readers' perception of pattern. Even an unretouched photograph is still composed and cropped to include or exclude detail that can change its message. Well-designed figures emphasize patterns the reader should see, but stop short of exaggerating patterns or implying them where none exist. This is something of a balancing act, and pulling it off requires some awareness of features of figure design that tend to mislead readers.

A lot of misleading design is rooted in scale. Whenever possible, vertical axes should start at zero, and when two panels or figures are to be compared, they should have the same scale (Box 12.5). Any deviation from this practice should be clearly signalled to the reader.

Fit lines are another easy way to mislead. The presence of a line through a scatterplot suggests a relationship, whether one is there or

Box 12.5 Misleading figure scales

When you ask a reader to compare panels within a figure, those panels should nearly always have identical scales. To illustrate this, imagine that you've measured biomass for plants grown in sun and shade, and with and without fertilizer. This is easy to show as a boxplot:

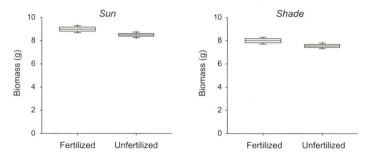

Notice that the panels' Y axes both begin at zero and have the same scale. It's easy to see that plants grew only slightly larger with fertilization and in sun.

It's tempting to think that much of the graph is wasted space—why not zoom in? Here the Y axes run from 7 to 9.5 g:

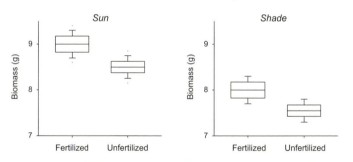

But the sun plot still has wasted space. You could start the shade plots at 7 g and the sun plots at 8 g.

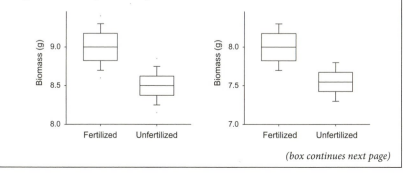

(box continues next page)

(cont.)

Here's the problem: each pair of plots shows exactly the same data, but they give very different impressions of the experiment's results. The second version greatly exaggerates the differences among treatments, while the third conceals the sun/shade difference and magnifies the fertilization effect. At best these require too much reader effort: only by mentally reconstructing the first version can readers accurately assess the pattern in the data. At worst, they're misleading. Many other features of figure design can similarly distort patterns.

not. Don't show a fit line if the fit isn't statistically significant (so there is no reason to suspect a relationship at all), and don't extend one beyond the range of the underlying data. Think carefully before using lines to connect data points, especially when data points are few: this gives the impression of much greater certainty about pattern than really exists.

Many additional problems lurk in figure design. Partly because the use of graphic design to mislead is a major pillar of the advertising industry, we know a lot about human perception and misperception of visual pattern (e.g., Ware 2012). You can—and should—exploit the same knowledge to tell your story without misleading.

I've offered only a few rules of thumb for the design of both figures and tables. Fortunately, all such graphics decisions come down to the usual question: what will favor crystal-clear (and accurate) communication with your reader?

Relating Graphics to Text

Figures and tables should both stand alone and—paradoxically—be seamlessly integrated with the text. That is, a reader, once directed to a graphic, should be able to understand it without referring back to the text. At the same time, the text should make clear what the reader is looking for in the graphic, and upon returning from graphic to text the reader should easily see how the graphic's content moves the argument along.

For figures and tables to stand alone, they need both good design (above) and also helpful legends. (For no apparent reason, tables have "titles" above them, while figures have "legends" below them. Name and position are the only differences.) A legend should begin with a brief phrase identifying the key point or comparison the graphic makes (for the figure in Box 12.4, for instance, "Biomasses of fertilized and unfertilized plants in sun and shade"). The remainder of the legend provides further explanation. It should define any symbols, abbreviations, or other coding of information not shown in an in-figure key or table footnote. In multipanel figures it should explain the panels and call attention to relationships between them. It should explain statistical methods used in the graphic (specifying, for instance, whether error bars indicate one or two standard errors, or what statistical methods underlie fit lines). It may include a brief reminder of the methods behind the data displayed—just enough for a reader to understand the data without having to look back to the Methods section. All of this should rarely take more than two or three sentences.

Do not expect readers to interpret a graphic unassisted. The text should indicate what pattern they should look for, how that pattern relates to the point being made, and how to see the pattern in a complex graphic. Avoid vague references to graphics, such as "See Table 1 for activation energies in the presence of different catalysts." Instead, first identify the pattern of interest, then direct your reader to the graphic that displays it: "Activation energies were lowest on palladium catalysts (Table 1)." When a graphic is complex, steer the reader as specifically as possible to the relevant features ("Figure 1, compare leftmost bars across panels"). Finally, avoid referring to more than one graphic to make a single point. I once read a manuscript that included this sentence: "Diet overlap between species increased from 2004 to 2009 in four of six comparisons: ribbon snake–green snake, mud snake–milk snake, milk snake–ribbon snake, and milk snake–green snake (Fig. 2A-F, Figs. 3–6, Table 3)." This asks the reader to do some difficult data-analytic work, extracting and synthesizing information from four figures and a table, which the writer should have done instead.

Chapter Summary

- Results are typically presented independently of the Methods and Discussion. However, they may include brief reminders of methods used and highlight results or comparisons for later discussion.
- Numbers are demanding for readers. Their impact can be minimized by winnowing data, avoiding redundancy, emphasizing the most important numbers, and displaying only meaningful and necessary precision.
- Tables are best for presenting datasets with many entries and/or variables, or when readers need precise numeric values. They are ineffective at showing trends or relationships between variables. Figures are excellent for highlighting trends and relationships, but do not display precise numbers or more than a few variables well.
- Figures may be data reproductions, schematics, or data compilations.
- Tables and figures should be designed to give readers the most information "in the shortest time with the least ink in the smallest space" (Tufte 2001, 51).
- Text should point readers to important pattern in tables and figures, but tables and figures should nonetheless be interpretable on their own.

Exercises

1. This small dataset shows harvest dry masses (in grams) for plants grown under four treatments. By hand or using your choice of software, construct Results section elements to present these data in the following ways. For each, write down a list of design choices you made, and why you made each one.
   a. Using only text, not a graphic.
   b. Using a table (of summary values, not just repeating the raw data). Make at least two different designs.
   c. Using a figure (data compilation). Make at least two different designs.

Which way of presenting the data is least, and which is most, effective? Why?

Watered daily to flow-through		Drought	
Fertilized	Unfertilized	Fertilized	Unfertilized
22	21	15	19
25	21	17	11
24	19	16	14
32	24	12	17
23	25	19	16

THIRTEEN

The Discussion Section

It's in the Discussion that writing practice varies the most, not just from discipline to discipline but even from paper to paper. The Discussion thus offers you great freedom in content and organization. For some writers, this freedom is exhilarating; for others, it's terrifying. For me, it's both.

Your Discussion's function is to turn data into knowledge. That means harnessing raw results to answer your paper's central research question(s) and to relate your answers to your broader discipline. The Discussion considers and extends your results to claim the strongest interpretation and the broadest importance that you can legitimately argue. All parts of that last sentence are important. It's entirely appropriate (and necessary) to blow your own horn by pointing out how your results provide novel understanding, resolve a longstanding dispute, or challenge or overturn previous knowledge—presuming, of course, that they do. It's also entirely appropriate to step beyond firm inference and speculate about what your results *might* mean. At the same time, reviewers and readers will notice, and resent, attempts to oversell results, to gloss over limitations in the data, or to misrepresent speculation as inference. Experienced writers avoid overstepping these rhetorical bounds with what linguists call "hedges" (Hyland 1998). Hedges are words or phrases that limit or fine-tune the strength of a claim: for instance, "It is *likely* that . . . ," "An *alternative interpretation* is that . . . ," "These data *suggest* . . . ," or "*If correct*, our model explains . . ." Of course, in order to hedge, you need to be aware of limitations or uncertainties in your argument. This means seeing your work as if you were a critical reviewer. This can be difficult (chapter 21), so it is very helpful, before tackling the

Discussion, to present your work at lab meetings or conferences where you can hope to be asked critical questions.

However it looks, your Discussion will be closely tied to your Introduction. The two are complementary: every question raised in the Introduction should be answered in the Discussion, and every major issue treated in the Discussion should be signalled in the Introduction.

The Elements of a Discussion

Although there is no universal recipe for constructing a Discussion, we can recognize four common components: interpretation of results, consideration of weaknesses, broader implications, and future prospects. These usually appear in that order—although some papers omit the weaknesses and/or prospects components—and they represent a gradual broadening of the paper's focus from the waist of the hourglass (Figure 13.1) out to its base. There's a lot of variation in the emphasis given the different components and in the rhetorical moves used to accomplish them (Swales 1990, 2004; Peacock 2002; Basturkmen 2012). Taking the four components in turn, and returning to our star-formation example, we can explore some common ingredients and some good practices:

- **Interpret results to answer your research question**. Because nearly all readers will attach importance to the first paragraph of a Discussion, this is the perfect place to supply a concise answer to your re-

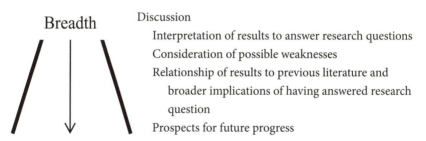

Breadth

Discussion
Interpretation of results to answer research questions
Consideration of possible weaknesses
Relationship of results to previous literature and
    broader implications of having answered research
    question
Prospects for future progress

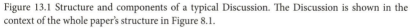

Figure 13.1 Structure and components of a typical Discussion. The Discussion is shown in the context of the whole paper's structure in Figure 8.1.

search question. For instance, either the first or the last sentence of our short opening paragraph might be "Our results imply that the cluster-assist mechanism is not necessary to explain massive-star formation in the Orion, Eagle, and Carina Nebulae."

Your Discussion will typically restate some of your results. But this does not mean simple repetition; use brief reminders and summarize rather than repeat in detail: "Solitary massive protostars were quite common in all three nebulae." The idea is to return the reader's attention to patterns pointed out in the Results, while considering what those patterns mean and how they answer the research question. It can be useful to complement this coarser-level summary of the results with specific examples that support explanations: "Furthermore, the most massive protostar observed (V3416) was one of the most isolated objects in our survey." However, remember that specific examples are just data points, and overdoing this means arguing from anecdote rather than analysis. Answering the research question may also draw on comparisons or synthesis of your current results with results from the literature.

A common rhetorical move is to comment on whether your results were expected or unexpected. In the latter case, you will usually identify what your expectation was, and where it came from: "Surprisingly, among massive protostars with close neighbors, the most massive did not have closer neighbors (on average) than moderately massive ones." Keep in mind, though, that surprises are likely to leave your readers unsatisfied unless you can also offer some explanation. The more support your explanation can have, the better—but it needn't be definitive and sometimes the best you can do is for it to be plausible. We might follow our protostar-surprise sentence with "This suggests that if cluster-assist does happen, it may only require a threshold level of gravitational assist to trigger a process that is subsequently independent of the neighboring masses. Theoretical models have not yet considered this possibility."

- **Consider possible weaknesses**. While there's likely to be hedging throughout the Discussion, it's often useful to include a paragraph or short subsection that explicitly addresses limitations to your inference. For instance, you might consider statistical power (which determines how the reader should interpret the *absence* of a looked-for pattern), weaker steps in a chain of logic, gaps in your data, or alternatives to your preferred interpretation of the results. We might, for instance, point out that our three nebulae aren't fully representative of all star-forming regions: "Our data leave open the possibility that the cluster-assist mechanism is more important in nebulae with higher gas densities; current instrumentation is unable to image any such nebulae with sufficient resolution."

Early-career writers sometimes greatly overdo this Discussion component, perhaps because we emphasize it when we teach undergraduates to write lab reports. If you leave your readers with the impression that your results don't mean much, they will feel they've wasted their time reading the manuscript.

- **Relate your results to previous literature, and considering broader implications**. While you may have called on data from the literature to help answer your precise research question (the first component), in this case I mean fitting that answer into its broader literature context. How does the answer to your research question relate to other research questions (answered or unanswered) in your discipline? Does it support current hypotheses in your field, or call them into question? Does it open up new questions? Does putting your work together with work from the literature establish consensus, draw attention to conflict, or advance our understanding in ways that individual papers can't? For instance, our star-formation results might be extended this way: "The relatively high frequency of solitary massive protostars complements the results of †Nasrallah et al. (2011)[1], who showed a high incidence of radiation jets from condensing protostars. Together, these lines of evidence suggest that the cluster-assist model is not necessary to explain massive protostar formation. They favor instead alternative models in which massive protostars form when gas clouds are heterogeneous in density, leading to asymmetric collapse and radiation escape in localized jets." With support from the literature, your paper's hourglass can regain the width it had at the beginning of the Introduction. Your ultimate aim is to show how the answer to your research question matters to the broadest possible set of readers.

---

[1] This reference is a fictional example, and throughout this book, I indicate such examples by the dagger (†) prefix.

- **Consider prospects for future progress**. Just as your work has inevitably built on past work by you and your colleagues, your results are likely to suggest directions for future work. For example, perhaps you have studied a specific case and your results point to a path to more general understanding: "While massive-star formation in the Orion, Eagle, and Carina Nebulae seems not to depend on the cluster-assist mechanism, future work should consider whether this conclusion can be generalized to nebulae differing in gas density and morphology. Planned enhancements to ALMA's resolution should soon make our approach feasible for a very large set of star-forming nebulae." Note that you should not simply offer the platitude that more research is necessary; instead, indicate specific directions such research should take. It may seem as if you're selling your own work short by admitting that it doesn't answer all questions. Not at all! In fact, you can emphasize the importance of your work in part by showing how it identifies future research directions that are likely to be profitable.

Do You Need a Conclusion?

One final issue bedevils writers of Discussions. Should your Discussion have a Conclusion? Of course you should have a small-c "conclusion": the answer to your research question. The question is whether this calls for a big-C "Conclusion" *section* (either a headed subsection at the end of the Discussion, or a separate section following it). There's no consensus on this question: practices vary among and within disciplines.

My own feeling is that separate Conclusion sections are seldom needed. When they exist, such sections usually do three things: reiterate the research question and its answer; state the significance or application of that answer; and identify directions for future research. But these are already typical elements of the Discussion. Unless your Discussion is very long and complex (or poorly organized), these points should be fresh in the reader's mind at its end. Because the last paragraph of your Discussion (like the first) is a power position—something readers will pay special attention to—this is a good place for a brief reminder of your most important finding and its significance. Extensive repetition to build a formal Conclusion section, however, is likely redundant. If your paper does seem to need a separate Conclusion section, before writing one, think about whether this is a sign that your Discussion needs to be simplified or reorganized. If your paper actually is complex enough that your reader can profit from a separate Conclusion, keep it short (perhaps four to six sentences) and avoid simply repeating the Discussion.

Chapter Summary

- The Discussion is the least formulaic of the major sections of a paper.
- The Discussion considers results and places them in context to claim the strongest interpretation and broadest importance you can legitimately argue.
- The Discussion complements the Introduction, dealing with every major issue raised there—but only those.
- Most Discussions (1) interpret results to answer the research question; (2) consider possible weaknesses of the study; (3) relate results to previous literature to suggest broad implications; and (4) consider prospects for future progress.
- Whether or not the Discussion ends with an explicitly labelled Conclusion, it should end with a brief reminder of your most important finding and its importance.

Exercises

1. Choose a recently published paper in your field, and read the Discussion. Highlight text comprising each of the four main Discussion components (interpreting results, considering weaknesses, relating results to literature, considering future progress).
2. What occupies the power position at the end of the Discussion you chose? What else might the author have placed there, and how might that choice have changed the paper?

FOURTEEN

||||||||||||||||||||||||||||||||||||||||||||||||||||||||||||||||||||||||||||||||||||||||||||||||||||||||||||||||||||||||||||||||||||||||

Back Matter

Acknowledgements

The Acknowledgements section is there to let you fulfill a social compact in the scientific community: that writers recognize contributions to their work. You might mention people who provided assistance with experiments, observations, or analyses; people who offered suggestions for your work or who commented on your manuscript; people who shared reagents, equipment, or other infrastructure; granting agencies or those who gave permission for your work (for instance, a government agency that issued you with collection permits).

Some readers enjoy Acknowledgements (I do). Others ignore them. But because the Acknowledgements have no real involvement with telling your story, they needn't keep to any particular conventions of style or content. Journals, as a result, tend to give you a free hand over this section.

References

The use of references, or citations, within the body of the paper is discussed in chapter 15. The References *section* is simply an organized list of works referred to in your text, and its function is twofold. First, it allows an interested reader to locate a work you cite in the literature. Second, it is the input to vast citation databases such as the Web of Science and Google Scholar. Through their linkage of citing and cited literature, these databases are increasingly used to measure the influence on scientific progress of papers, authors, journals, academic units, and so on. Citation-based measures of impact are demonstrably imperfect (e.g.,

Ramsden 2009, Lozano et al. 2012), but this is of little immediate importance, because such measures aren't going to be abandoned any time soon.

The reader-directing function of this section is assisted by, and the databasing function requires, very strict consistency in the presentation of your reference list. Unfortunately, journals vary considerably in required format; but fortunately, software is widely available to make compliance straightforward. Popular alternatives at the time of writing include Mendeley (http://www.mendeley.com), Reference Manager (http://www.refman.com), JabRef (http://jabref.sourceforge.net), and RefWorks (http://www.refworks.com).

Appendices and Online Supplements

It has always been possible for scientific papers to include appendices, which typically consist of material that supports your paper's story (chapter 7), but isn't necessary to allow most readers to understand it. When journals existed only in print, appendices were uncommon and typically short, because they consumed limited resources (journal space and editorial attention). Printed appendices have now been replaced almost entirely by supplements accessed online. Because digital appendices are enormously cheaper to publish than printed versions, the move online has dramatically changed writing practice. Journals are willing to publish many more and longer supplements (especially if doing so can shorten the paper's main body), and writers seem to love providing them. In fact, online supplements can seem like candy to a writer: no hard decisions are needed, because everything can be included! Resist this temptation. Supplements are cheap, but they aren't free, and their costs crop up throughout the lifecycle of a scientific paper. The time it takes to produce and polish supplements might be better spent on improving a manuscript's main body (or starting another one). Easy recourse to supplements can also weaken your commitment to finding and telling a well-defined story. During the publication process, excessive supplements increase demands on reviewers and editors. For readers, supplements present some costs of navigation (as they leave the main text, find the supplement, and then return). These costs are minimal for online readers of journals that provide direct hyperlinks to the supplements; they are much more substantial when such hyperlinks aren't available, or for those who read offline. Finally, readers faced with long lists of supplemental material are likely to shrug and ignore them all: if everything is important, then nothing is. For all these reasons, think carefully about which supplements are truly necessary.

How do you decide whether a particular piece of information belongs in the main body of your paper, or in a supplement, or shouldn't be published at all? In general:

- If readers need that piece of information to understand the story you are telling, it belongs in your paper's body.
- If readers don't need the information to understand your story, but some will find it useful for other reasons, it's a good candidate for a supplement.
- If no plausibly imaginable reader will use the information, then it shouldn't be published at all, either in body or supplement. I say "plausibly imaginable" because you can't see the future, but you also can't include everything just in case. So if you can't explain what kind of reader might use the information and why, leave it out.

Little other than your imagination limits the kinds of material that can appear in supplements. Common supplemental material includes:

- Mathematical derivations or proofs (e.g., Heard 1992). Of course, the proof itself may be the story of a paper, particularly in mathematics; but if the proof's existence is important to the story, but its nature is not, then it belongs in a supplement.
- Methodological details, which may describe procedures (e.g., Soutullo et al. 2005), computations (e.g., Halverson et al. 2008b), field site locations (e.g., Kolaczan et al. 2009), or the like. A synthetic chemist might include detailed experimental conditions; a particle physicist might include schematics of a detector. Supplements are useful for information needed by readers who want to use the same methods, but not to readers who want only to understand the current result.
- Extra figures or analyses (e.g., Woods et al. 2012). These will generally support a point made in your paper, but one not central to the main story and that most readers would be willing to take your word for.

Such supplements increasingly replace the older, and rather unsatisfying, custom of mentioning "data not shown" to support such a point.

- Samples of data types not easily reproducible in print, such as high-resolution imagery, sound, or video (e.g., Carter and Wilkinson 2013). Note that the intent here is not to provide all the raw data, but rather to illustrate the nature of the data so a reader can understand how the work was done.

- Detailed datasets (e.g., Nason et al. 2002). Data supplements are appropriate when the data have value beyond the results that are part of the paper's story. Presenting all the data in the paper's main body would impede the telling of its story; but doing so in a supplement allows interested readers to reanalyze the data for their own purposes. For instance, in organic chemistry it's routine to provide NMR spectra for all newly synthesized compounds, so that anyone later using the synthesis methods can check their products' purity and nature by comparing spectra. This type of supplement is rapidly being replaced by the data archive (see next section).

- Software written for use in your analyses. This may consist of short pieces of code run under a larger package available to your readers, such as Excel macros or R scripts (e.g., Oke et al. 2014), or longer programs written as standalone executables (e.g., Heard and Cox 2007). While some journals insist on submission of custom software as part of the review process, it's rather unlikely that reviewers or readers will inspect the software to see if it works as claimed. The real reason to provide it is to make it available for use by readers. Journals differ as to whether software should be provided as a supplement via the journal's own website or posted to a software-distribution site such as Github (http://www.github.com). The former provides the most stable archive and assures the reader of retrieving exactly the software you used. The latter allows you to update your posted software and encourages its further development by others, but it's not clear whether any particular software site will still be around when a future reader wants to retrieve your code. It's probably best to post software both ways.

When you submit your paper, you're likely to find supplements bound by less exacting format requirements, and receiving much less rigorous attention from reviewers and editors, than your paper's main body. This does not mean that you should pay them less attention yourself, or see them as a route to get weaker inferences past the scrutiny of reviewers. If anything, it means you should take supplements *more* seriously, because you shouldn't expect much help from peer reviewers in identifying weaknesses or errors in them. If you are unconcerned about problems creeping into a supplement, it probably isn't needed in the first place.

Data Archives

The ease and low cost of online data storage and retrieval has led to the rapid spread of policies that encourage, or even require, the public (online) archiving of all data analyzed in a scientific publication. Such archives are valuable for many reasons, most obviously that they ensure the availability of data for future meta-analysis or reanalysis for purposes not conceived of by the original authors (Whitlock et al. 2010). The archived data do not appear with the published paper and are generally not part of the manuscript as prepared for journal submission. What appears in the manuscript is simply a pointer (web link or otherwise) to the actual location(s) of the data archive(s).

Some kinds of data have their own dedicated archives: for example, DNA sequence data are usually deposited in GenBank (http://www.ncbi.nlm.nih.gov/genbank/), chemical crystal-structure data in the Cambridge Structural Database (http://www.ccdc.cam.ac.uk), and measurements of Earth's magnetic field in SuperMAG (http://supermag.jhuapl.edu). Other kinds of data can be deposited in general-purpose archives such as Dryad (http://www.datadryad.org). In many cases, deposited data can be "embargoed," with public access temporarily blocked, so that scientists can have exclusive use of their own data for a reasonable period.

Archiving policies vary among fields, journals, and archives, as do definitions of "primary data" to be archived (original video footage? measurements derived from that footage? something else?) and policies controlling public access to and use of archived data. It is wise to investigate the archiving policy of your target journal early in the writing process, because it is far easier to prepare data archives during data analysis than it is to reconstruct datasets months or years later.

Chapter Summary

- "Back matter" includes acknowledgements, references, appendices and online supplements, and data archives.
- Acknowledgements fulfill a social compact that we recognize contributions to our work.
- The Reference list follows a strict format that varies by journal; software makes this formatting straightforward.
- Appendices and online supplements include material that supports the story your paper is telling, but isn't necessary for every reader to understand that story. Content often included in supplements includes derivations or proofs, methodological details, extra figures or analyses, samples of data, and detailed datasets.
- Archiving of raw data is now widely required as a condition of publication.

Exercises

1. Choose a recently published paper in your field that includes online supplements. What material appears in each supplement? What value will the content of each supplement have, and to what kind of reader? For each supplement, why was the content not included in the main paper?

FIFTEEN

Citations

Science is cumulative: new research builds on earlier work, extending, reinterpreting, or correcting it. As a result, every modern scientific paper is studded with citations to the literature. Good writing uses in-text citations to strengthen storytelling, but deciding which and how many sources to cite depends on some understanding of the function of individual citations and of citation more generally.

Citation as a Four-Party Transaction

Every citation is a transaction involving four parties: writer, reader, publisher, and source (cited author). Understanding citation practices requires some thought about the functions, benefits, and costs of citation for each.

At the simplest level, citations pass information from writer to reader. A citation may establish what is known, helping define your research territory; or it may establish what is *not* known, helping establish your niche within that territory. It may provide methodological detail (keeping your Methods section concise and readable). It may support interpretation of your results or establish their importance (for instance, citing similar findings or the earlier theory that your results test).

More broadly, citations assist with communication because they help establish your authority. Readers trust claims that are supported by appropriate citations, and writers who show by citations that they understand the context of their research question and know what other authors have said about it.

If citation practices had to satisfy only writers and readers, they would be fairly simple: writers would use enough citations to support claims and establish authority, but no more. However, considering the interests of publishers and sources adds some complexity, because these parties see citation differently from writer and reader. The result can be some disagreement over optimal citation practice.

For publishers, citations carry costs of editing, typesetting, proofing, and printing, but few benefits. Some publishers explicitly or implicitly encourage citations to recent papers in their own journals (which increase those journals' impact factors, the primary tool by which they market themselves to contributors and subscribers). Otherwise, publishers tend to press writers to reduce citation. They may set a maximum allowable number of citations (e.g., *Nature*), specify use of a "reasonable minimum" (*Endocrinology*), or simply count the reference list toward a limit on overall manuscript length (*Geology*). Many journals particularly target citations in the Introduction, resisting the provision of comprehensive literature reviews.

Source authors, by contrast, love citations. A scholar who cites your work acknowledges your contribution to the progress of science and your place in the scientific community. The ever-growing importance of citation-rate data to assessments for hiring, promotion, tenure, and grants makes the benefit to source authors an unavoidable part of our thinking about citation. There is something of a social compact among scientists: I cite your work in part because I would very much like you to cite mine.

How Many Citations to Use

Thinking about citations as four-party transactions helps explain why writers are sometimes uncertain about how many to use. The writer and reader's agreement on the minimum needed to make the argument convincing becomes muddled by pressure from publishers to reduce citations and from the social compact to increase them. Getting the number of citations right means balancing all these forces.

When should a statement or claim be supported by citation? A statement that's widely accepted doesn't require any citation at all: for instance, that stars exist in equilibrium between gravity and radiation pressure (Box 7.1). Neither does one that is factual and easily checked (that milkweed is a perennial plant in the family Apocynaceae). Similarly, methods that are standard in a field can be mentioned without citation (NMR spectroscopy in organic chemistry). Citations *should* be supplied for statements that readers might question, for methods that are unfamiliar, or when there is potential value in referring interested readers to more information. These are judgement calls and conventions vary among fields. Only through familiarity with recent literature in your own field can you be sure what needs citation and what does not.

When citation is required to support a claim, many writers are tempted to pile on, reasoning that if one citation is good, several are better. Resist this temptation (outside of a review paper, where comprehensive coverage of the literature may be the point). Readers will appreciate your effort in choosing the citations that are most relevant and useful. Nearly always, one to three citations can support a point; more carry little additional information but make your text difficult to read. Even if your claim is of the form "everyone believes thing X," don't cite everyone! You can use "e.g." to be unambiguous that you are providing only representative citations: "(e.g., †Xi 2007; †Jones 2009)." Or you might summarize literature by citing a recent review: "(review: †Schmidt 2012)."

What of the total number of citations in a manuscript? Because conventions for what requires citation vary among fields, you might expect the length of citation lists to vary too. Surprisingly, though, the Web of Science suggests a remarkable consensus, with twenty-five to sixty citations typical for major research articles in most fields (e.g., ecology, cell biology, organic and physical chemistry, astronomy, geoscience, and condensed-matter and particle physics). Pure math papers are the major outlier, with six to twenty citations being common. Short "notes" and commentaries will have many fewer, and review papers many more (well into the hundreds). If your manuscript is outside the typical range, ask yourself why.

Which Sources to Cite

It's not unusual to find that you could support a particular claim by citing any of a dozen different sources. In choosing among them, ask yourself only one question: which citation(s) will be of greatest use to the reader who wants to learn more?

Cite the most relevant sources, avoiding material that supports your point indirectly or only when several papers are considered together. While it may be clever of you to notice subtle connections between your work and superficially unrelated literature, unless that connection is the only path to insight the citation will only impede communication with your reader.

Given several options of similar relevance, publication date and accessibility are other considerations. If you are citing to give credit (for an idea, model, method, etc.), you should cite the earliest source, but if you are citing to establish current understanding of something, you should cite the most recent. You should also choose sources that a reader can access easily. This means citing, when possible, a journal paper over a book chapter, thesis, conference abstract, or technical report. Cite well-known and widely held journals over obscure or regional ones. Many scientists prefer to cite open-access journals over subscription ones.

Finally, avoid citing secondary sources (review papers or books that summarize and synthesize the primary literature) if citing the original source will serve. Cite a review, instead, in reference to its synthetic findings—a pattern, for instance, evident only from surveying many publications. Citing a review for a particular result mentioned therein shunts the work of tracking down the original literature onto your reader, and deprives the original author of a citation. In the rare case—and it should be very rare—that you cannot track down the original, cite it nevertheless, but admit that you haven't seen it: "(†Smith 1907, as cited in †Singh 2006)."

Odd Kinds of "Citations"

While the majority of citations refer to formally published works, you will occasionally use an odd beast: a "citation" of something unpublished. These include:

- "(pers. comm.)," or "personal communication." This means "somebody who ought to know told me." The source should be identified as clearly as possible, with initials or full name or, better, with an academic or other affiliation ("†A. N. Mbala, Springfield University, pers. comm."). Editors often ask for evidence that such citation is appropriate, such a letter from the cited person approving the exact wording.

- "(pers. obs.)," or "personal observation." This means "I noticed this, but don't have any formal data." Such citations should never be used for claims that are central to a paper's argument or likely to be contentious, but they are sometimes used for ancillary or background information. For example, in Heard and Kitts (2012) I wrote "*Solidago gigantea* and *S. altissima* [plants] . . . spread clonally, with [genetic individuals] . . . often including very large numbers of [stems] (Maddox et al. 1989; S. Heard, pers. obs.)." The Maddox paper described only the growth form of *S. altissima*, so it would be misleading to cite only that paper; and no citation was available for *S. gigantea*. We decided that a *pers. obs.* citation would suffice because readers were unlikely to doubt the description. Arguably, on the same grounds perhaps we needed no citation at all.

- "(unpubl. MS)," or "unpublished manuscript." This refers to a manuscript not yet accepted for publication; when it is accepted it becomes "in press" and can be cited normally. Many journals disallow such citations, reasoning that until a manuscript has passed muster in peer review, it has no more claim to authority than the *pers. obs.* (for your own manuscript) or *pers. comm.* (for somebody else's) that could serve instead. I disagree, because such a citation lets an interested reader contact you to ask for the manuscript, or search to see if it was published since you made the citation. This only works, though, for manuscripts complete enough that they are likely to be published in approximately the form cited. A useful rule of thumb might be to cite only manuscripts that have actually been submitted for review.

- "(unpubl. data)" or "(results not shown)." These suggest that you can support your claim with data or analyses that you are not including in your manuscript. The aim is usually either to reserve data for a future paper to which they are more integral, or to remove less important data or analyses to reduce the burden on the reader. These citations can also be used to indicate that alternative methods yield consistent results: for instance, you might write "We compared ejecta volumes across volcano types using parametric ANOVA; however, nonparametric and randomization tests supported identical interpretations (results not shown)." The rationale here is that including multiple analyses would expand the paper without much improving the argument. This kind of citation is being replaced by online supplements (chapter 14) as they become cheap and easy vehicles for ancillary data and results.

These odd citations are uncommon, and for good reason. They provide only weak support for a claim: at best, they are the writer's parenthetical "no, really!" They should never be used to support an important claim, and if a claim is unimportant enough for weak support to suffice, one might wonder whether it belongs in your story in the first place. Deploy these citation types only when no better alternative exists.

Chapter Summary

- Citations involve four parties: writers, publishers, readers, and sources.
- Citations have functional roles (supporting your argument) but also help establish your authority and reward source authors for their contributions.
- Citations are needed for claims that may be questioned, for unfamiliar methods, or when readers may want additional information. Common-knowledge facts and standard methods don't need citations.
- "Pers. comm.," "pers. obs.," "unpubl. MS," and "unpubl. data" citations should be minimized, but are occasionally useful.

Exercises

1. Choose a recently published paper in your field and read the Introduction. Identify three claims or statements that are supported by citations. What functions do those citations play for the writer and for the reader? Identify three claims or statements that are *not* supported by citations. Why not?

SIXTEEN

||||||||||||||||||||||||||||||||||||||||||||||||||||||||||||||||||||||||||||||||||||||||||||||||||||||||||||||||||||||||||||

Deviations from the IMRaD Canon

Having genuflected at length before the canonical IMRaD structure, I must now acknowledge that it doesn't work for every scientific paper. Over your career, you will encounter—and write—papers that use a variety of noncanonical structures. (Writing forms other than the scientific paper are treated in chapter 25.) Some alternative structures merely bend the IMRaD canon slightly, such as the placement of Methods at the end of the paper. Others, such as review papers, appear to discard the canon entirely.

Escape from the canonical structure may seem freeing, but the canon is there for good reason (chapter 8). You should, therefore, deviate from IMRaD not when doing so seems to make the paper easier to write, but only when you are sure it makes the paper easier to read. If you are forced to deviate (as for Notes, for example; see below), do so as mildly as possible.

"Methods-Last" Papers

An alternative structure that's becoming more common—*Nature* and *Cell*, for instance, require it—is placement of the Methods at the end of the paper. On the surface, this seems bizarre: how can it make sense to first report your Results, and only later describe the experiments they came from? But on closer examination, methods-last papers aren't really so exotic. While the *section* labelled "Methods" is placed at the end of the paper, plenty of information about the methods is supplied in roughly the usual position. The last paragraph or two of the Introduction usually indicate the overall approach, and the Results tend to be full of methods-rich sentences such as "To compare the relative tumorigenic potentials of these [cell] subpopulations, we purified [cells] by fluorescence-activated cell sorting . . . [These cells] were injected immediately into the mammary fat pads of NOD/SCID mice" (Chaffer et al. 2013). In an IMRaD paper, these sentences wouldn't belong in the Results; but in a Methods-last paper they're needed there, because otherwise the reader would have no hope of understanding the tumor-growth data to be presented.

If you write for a Methods-last journal, think about this structure as recognizing two classes of readers. Most need just enough information about your methods to understand the results you report; for these, you slip summaries of the methods into Introduction and Results. A few readers want additional detail (perhaps they will borrow your techniques), and only these will bother with the Methods section at the end of your paper. Thus, Methods-last papers are really just near-conventional IMRaD papers that reserve the label "Methods" for an Appendix holding supplemental methodological details.

Review Papers

The review paper is the non-IMRaD form that writers (and readers) deal with most often. A review answers a research question by drawing on many investigations—by surveying and evaluating the literature on a topic. Reviews are usually less detailed and technical than primary-results papers, because they're typically intended for a broader audience, including nonspecialists and researchers curious about, or intending to enter, the field. Finally, reviews ask research questions broader in scope. For example, one of my students published a primary-results paper asking the specific question "How does (simulated) insect herbivory affect mortality and seed production of the rare aster *Symphiotrichum laurentianum*?" (Ancheta et al. 2010), and then a review asking the much more general question "How does insect herbivory affect the population biology of rare plants?" (Ancheta and Heard 2011).

Review papers retain two features of IMRaD structure: opening with a general Introduction, which makes clear the review's subject and scope by identifying the research question(s), and closing with some kind of a conclusion, which summarizes its answer. Beyond this, reviews share no standard organization. The main body is nearly always divided into sections, but these are often more akin to subsections of a Discussion than they are to the main IMRaD sections. Common choices for organizing reviews include systems of temporal, methodological, or thematic headings. Temporal organization retraces the historical development of knowledge on the topic, which is effective only when the research question directly concerns that history. Methodological organization is common when a review synthesizes theoretical and empirical knowledge. For instance, a review of massive-star formation might begin with a section covering theoretical models of gas-cloud collapse, and then continue with sections summarizing telescopic data from different instruments or wavelengths available to test those models. Thematic organization is the most flexible and most common; here sections are arranged by topic in a logical order to draw the reader through the review while maintaining focus on the research question.

My student's rare-plant herbivory review (Ancheta and Heard 2011) used a topical organization within an IMRaD skeleton:

1. Introduction
2. Methods
3. Results and Discussion
    3.1 Herbivore impacts on rare plant populations: quantity and quality of data
    3.2 Documented impacts of herbivores on rare plant populations
    3.3 Density-dependence
    3.4 Biocontrol herbivores
    3.5 Towards a more sophisticated understanding of insect herbivory on rare plants
4. Conclusions

In the Introduction, we set up our research question, "How does insect herbivory affect the population biology of rare plants?" and placed it in literature context. Our Methods section describes how we located relevant studies and extracted herbivore-impact data from them. Many reviews omit Methods, but for "quantitative" reviews that analyze data extracted from multiple publications, these are important. Such reviews have become common, as formal meta-analysis (Borenstein et al. 2009) is increasingly applied to the synthesis of published results.

Our main section ("Results and Discussion") has five subheads. It begins (3.1) with an overview establishing the scarcity of high-quality data bearing directly on the research question. Because of this scarcity, we next (3.2) tabulate studies that help answer only the simplest version of our research question: can insect herbivory reduce survivorship or reproduction of rare plants? Having answered that question (positively), we turn (3.3) to a more sophisticated, but harder to answer question: can insect attack depend on plant population density in a way that stabilizes plant populations? Section 3.4 identifies an important narrower question: can insects introduced for weed control have unintended effects on rare native plants? We then (3.5) offer some recommendations for future studies. Finally, our brief Conclusion reiterates our research question, its significance, and the best answer we can supply.

There were many other ways we might have organized our review, and we tried several of them during writing. We settled on the organization above because it led logically toward a (provisional) answer to our research question, and also did a good job of highlighting the severe shortage of the kind of data needed to make our answer more certain. Because reviews offer many more organizational options than other kinds of papers, outlining and related techniques for finding your story (chapter 7) are especially valuable tools as you write them.

Mathematical and Theory Papers

Mathematical papers, or theory papers using mathematics in other disciplines, are sometimes treated as if they were very different from observational or experimental papers. Some are. Perhaps the furthest removed are papers in pure mathematics, many of which dispense almost entirely with introductory material and discussion. Such papers begin with the statement of a theorem and end with the completion of its proof—and thus seem to consist entirely of Results. A wealth of detail particular to writing in mathematics is provided by Higham (1998). Even outside mathematics, many writers of theory papers find IMRaD structure unsuitable, with effective organization having little to do with the conventional divisions among Methods, Results, and Discussion. But other writers of theory papers adopt more or less recognizable versions of IMRaD structure, even if they don't use quite the same headings. I fall into this camp: among my own theoretical biology papers, Heard and Remer (2008) uses completely canonical IMRaD structure; Heard (1995) adds an extra section, "The Model," between Introduction and Methods; and Heard and Remer (1997) uses "Model and Results" in place of separate Methods and Results sections but is otherwise IMRaD.

That IMRaD structure can often accommodate mathematics or theory should not be too surprising. Theory papers should define the problem to be solved and outline its context and importance: an Introduction. They should establish the mathematical techniques used to construct equations or models, and to solve them (analytically, numerically, or by Monte Carlo simulation): a Methods section. The model's solution (or data from simulations) must be reported: the Results. Finally, theory papers should leave their readers with an appreciation for how their results advance our knowledge of the field: a Discussion. All the IMRaD sections are there. The only question is whether deviations from the canon can improve communication with the reader, and if so, whether those deviations should be mild or more marked. For instance, integrating the setup ("methods") and solution ("results") of a model might allow smoother flow; and renaming sections (for instance, "The Model" instead of "Methods") can provide the reader with a clearer finding system. Because IMRaD offers readers so much, I believe that deviations from it should be as mild as possible. Depart further only if you are genuinely convinced that doing so greatly advantages the reader.

Descriptive Papers

In the early years of the modern scientific journal, a great many papers were simply descriptions of novel specimens or unusual events[1]. Such papers have gradually declined in frequency, but they remain important

---

[1] Sometimes very unusual, as in Robert Boyle's (1665a) *An Account of a Very Odd Monstrous Calf.* The calf had deformed legs, a divided tongue, and no obvious scientific significance. Boyle followed this up with a description of a deformed colt entitled *Observables upon a Monstrous Head*

in some disciplines, such as biosystematics (e.g., descriptions of newly discovered species), earth science (e.g., descriptions of stratigraphy or mapping of faults), and astronomy (e.g., descriptions of extrasolar planets or other astronomical objects). These modern descriptive papers are much more sophisticated than their ancestors, of course, but they retain the function of describing nature rather than necessarily, or directly, testing hypotheses about its function or origin.

Descriptive papers can deviate from IMRaD, although as for theoretical papers, this deviation should be as modest as you can manage. Descriptive work still has context, methods (by which you collected samples or made observations), results (the description itself), and implications or significance—all the conventional elements of IMRaD structure. Most descriptive papers essentially adopt the canonical structure but rename a section or two. For instance, Li and Qian's (2013) description of binary stars in the globular cluster ω Centauri renamed the Methods "Light Curve Analysis of the Two Binaries," while Gulick et al's (2013) seismic imaging of an Alaska fault system renamed the Results "Observations and Interpretations." Both papers were otherwise IMRaD. Other papers, such as Hind and Saunders (2013), accommodate description within conventionally headed IMRaD sections. While some descriptive papers abandon IMRaD entirely, these examples illustrate that they need not do so.

Notes

Very short papers are often treated distinctly by journals as "Notes." Probably just to save space, many journals prohibit division of Notes into headed sections. The very shortest papers of this form (such as Saunders and Clayden 2010, a half-page correction to an earlier paper) lack much introductory material or discussion and so need little in the way of structure. Otherwise, the fact that IMRaD *headings* aren't allowed should not prevent you from taking advantage of IMRaD *organization* to help guide readers through your Note's logic.

(Boyle 1665b). Boyle is celebrated today for his contributions to physics and chemistry, but not biology.

Comments

"Comment" papers critique or attempt to rebut recently published primary-results papers. These deviate more routinely, and more extremely, from IMRaD structure than any other kind of paper. Because Comments are not intended to be read independently of the paper they comment on, they usually have little introductory material—at most, a few sentences summarizing the original paper and explaining why debate about its interpretation matters. Most Comments don't have methods or results to report; instead, they offer perspective on the original paper's methods and results. That leaves Discussion; and in fact, a Comment is perhaps best understood as an alternative or supplement to the original paper's Discussion. Organizing a Comment is thus more like organizing a Discussion than organizing a whole paper.

Actually, structure and organization are the easy part of writing a Comment. Far more difficult is writing a Comment that critiques the original paper's science without attacking its authors. Great care is needed in choosing wording and balancing criticism with respect. (Compare Janssen [2013], which does this well, with Bizarro [2013], which does not.) Montgomery (2003:109ff) provides useful guidance.

Chapter Summary

- Not every paper adopts the IMRaD structure.
- "Methods-last" papers need some information about the methods to appear in Introduction and Results.
- Review papers retain some IMRaD elements, including Introduction and general Discussion, but otherwise can be organized in many ways.
- Mathematical, theoretical, and descriptive papers sometimes deviate strongly from IMRaD, but they need not; all these include elements recognizable as introduction, methods, results, and discussion.
- Short "Notes" seldom use IMRaD headings, but should retain IMRaD structure.
- "Comment" papers are best understood as alternatives or supplements to the Discussion of the paper they comment on.

Exercises

1. Choose a recently published paper in your field that does not use IMRaD section headings. Highlight material you think plays the functions of each IMRaD section. How does the overall organization differ from IMRaD? Why do you think the authors chose that organization?

Part IV

||||||||||||||||||||||||||

Style

Just as the scientific paper is built from the sections that were the focus of Part III, sections are built of small units combined into larger ones: words put together to build sentences, and sentences to build paragraphs. Clear paragraphs and grammatical sentences populated with appropriately chosen words are tools you use to achieve clear communication with your reader.

There is a difference, though, between structure and composition at the section scale and at the words-to-paragraphs scale (I'll call the latter "style"). Sectioning of a scientific paper involves conventions that are particular to writing in the sciences, but many (not all) issues of paragraph structure, sentence structure, and word choice are universal. For instance, rules of grammar, devices we use to link one paragraph to the next, and the dangers of troublesome word pairs such as *affect/effect* are the same no matter what we write about. This means that skills you've built in writing good sentences and paragraphs of any kind transfer with only minor adjustments (I'll mention a few) to scientific writing. It also means that general guides to English composition and style are perfectly suitable for consultation by the scientific writer. My own coverage of style doesn't pretend to be exhaustive. (Among more comprehensive treatments, see Williams 1990, Fowler and Aaron 2011, and Sword 2012.) Instead, I emphasize some particularly important principles, explore a few areas in which scientific writing poses its own challenges, and cover some points of style that seem especially problematic for scientific writers. I'll begin by dissecting sections into paragraphs, zoom in to consider sentences and then words, and finally discuss the important issue of brevity. I'll come at this material from the perspective of *scien-*

tific writing, and in particular the journal paper, but of course most of what you learn here will transfer to writing of almost any other kind.

Scientific writing isn't art (although it needn't be ugly, either; see chapter 28), and nobody expects your readers to swoon over the lyricism of your prose. Nevertheless, attention to style can make your writing more effective, and your reader grateful.

SEVENTEEN

||||||||||||||||||||||||||||||||||||||||||||||||||||||||||||||||||||||||||||||||||||||||||||||||||||||||||||||||||||||||||||||||||||||||||||||||||||||||||||

Paragraphs

The most obvious unit of organization within sections is the paragraph. Paragraphs are obvious because we use line breaks, indentation, and white space to make breaks between them that can't be missed. All this typographic attention suggests that organization of material into paragraphs must be really important, and that's true. Good paragraph structure can make communication with your reader enormously clearer. This is especially true for scientific writing. True, scientific paragraphs look and function just like any other paragraphs—but because we write about material that's inherently complex and highly technical, the stakes are arguably higher than for other kinds of writing. The major focus of this chapter is on how you can use paragraph structure to make complex material as effortlessly clear to your reader as possible.

You probably decided on the rough ordering of material you need to package into paragraphs during your story-finding steps (chapter 7). What's next is to tackle actually writing those paragraphs, and connecting them so they work together.

The Nature of the Paragraph

A paragraph is a unit of logical organization best thought of as introducing and treating a single idea. Thus, each paragraph break marks the appearance of a new idea in your argument. (In this chapter's examples, I emphasize paragraph breaks by including the paragraph symbol, or pilcrow: "¶.") What constitutes a "new idea" meriting a new paragraph can be subtle, though. A paragraph may introduce an idea that's entirely new, or it may take a new perspective on something already under discussion (for instance, shifting from advantages to disadvantages of a procedure, or from evidence supporting a hypothesis to that opposing it). It may mark movement through a sequence of related ideas:

¶Three features of the data suggest . . . First, . . .
¶Second, . . .
¶Third, . . .

Occasionally, you might even engineer an extra idea shift deliberately, so that you can give your reader a paragraph break as a breather during what would otherwise be an overlong passage. How long is "overlong"? There is, I'm afraid, no simple answer, beyond "too long for effortless and crystal-clear comprehension by the reader." Experience (and reviewers) must be your guide.

Three Properties of Good Paragraphs

Good paragraphs have three important properties: they are unified, coherent, and distinct. All sentences in a "unified" paragraph deal with the same idea. A "coherent" paragraph has sentences that work together effectively to develop its idea. A "distinct" paragraph is self-contained, differs in topic from paragraphs preceding and following, and communicates its idea without strong dependence on other paragraphs.

Making paragraphs unified. If your paragraph is to be unified around a single idea, you need to know, and tell your reader, what that idea is. Clearly declaring the paragraph's central idea is the job of its topic sentence. (Actually, a paragraph may have topic *sentences*: sometimes you may take two or three—but never more—to declare a topic. As you read here, just tack a mental "(s)" onto "topic sentence.")

In scientific writing, the topic sentence is nearly always the first in the paragraph. This is one of a paragraph's two "power positions," where readers expect to find important information. The other is its *last* sentence, a good place for a succinct statement of the paragraph's take-home message. When you structure a paragraph this way, you announce your destination right up front, bringing the reader along for a smooth ride to the expected finish.

A topic sentence declares the paragraph's central idea, but does not have to fully explain it. This is a good thing, because in scientific writing a paragraph's idea will often be complex enough that your reader won't fully understand it without the logical development to come. Such situations can be accommodated with a topic sentence along the lines of "¶We can evaluate the cluster-assist model by considering three features of our protostar mass and distribution data." This provides a road map to the paragraph—not with every detail, but with enough about its route to avoid any major surprises.

With the topic sentence out of the way, the body of the paragraph is devoted to development and support of the central idea, including logic, details, examples, and so on. *Every* sentence in your paragraph should contribute to developing the idea announced by the topic sentence. Deviations from this leave your readers feeling a kind of mental whiplash, wondering whether they have missed an intended connection or misunderstood the paragraph's announced idea. For example, imagine that you read this:

¶[(1)]Two main forces are at work as protostars accrete matter from clouds of dust and gas: gravity and radiation pressure. [(2)]The force of gravity drives accretion, and increases as more material is accreted. [(3)]Radiation pressure opposes accretion, and increases as gravitational collapse increases the protostar's temperature. [(4)]As accretion progresses, radiation pressure strengthens relative to gravity, and accretion should cease when the two forces balance. [(5)]Electromagnetic forces drive stellar winds. [(6)]The simplest models suggest that gravity-radiation equilibrium should occur before the protostar reaches about 10 solar masses. [(7)]The existence of larger stars, therefore, suggests that these models are incomplete. [(8)]Very massive stars are the source of elements heavier than iron in our universe.

This paragraph probably leaves you unsatisfied. Sentence 1 (the topic sentence) declares that the paragraph is about the two forces that control accretion of mass to a protostar. All seems well as these forces are identified and sentences 2–4 explain their roles. Sentence 5, though, mentions a third force, one missing from the topic sentence. Are electromagnetic forces related to protostar accretion in a way you haven't grasped? Or is the topic of the paragraph not "forces controlling accretion" after all? Or does the sentence simply not belong? Sentences 6–7 return to the declared topic, but sentence 8 drifts away again. The writer (me)[1] should have recognized sentence 5 as a stray thought and deleted it, and realized sentence 8 was better placed in an earlier paragraph justifying interest in massive stars. Read the paragraph again without them, and notice the difference:

¶Two main forces are at work as protostars accrete matter from clouds of dust and gas: gravity and radiation pressure. The force of gravity drives accretion, and increases as more material is accreted. Radiation pressure opposes accretion, and increases as gravitational collapse increases the protostar's temperature. As accretion progresses, radiation pressure strengthens relative to gravity, and accretion should cease when the two forces balance. The simplest models suggest that gravity-radiation equilibrium should occur before the protostar reaches about 10 solar masses. The existence of larger stars, therefore, suggests that these models are incomplete.

This version passes the "unity" test: each sentence relates clearly to the idea declared in the topic sentence.

Making paragraphs coherent. If the sentences that make up a unified paragraph are to work together to develop its idea, they need to present material in a logical order, and to relate each step of idea development to the next. These functions are accomplished by paragraph organization and by relational devices.

In a well-organized paragraph, material is arranged to give your reader a mental framework on which to hang all the information the paragraph offers, and so that whatever information the reader needs to understand any given sentence has already appeared in previous ones. There are two major techniques you can use to achieve these goals: signalling in the topic sentence, and adoption of a standard organizational scheme.

---

[1] You might think it's difficult for me to write plausible but deliberately flawed paragraphs to serve as examples. If so, you're confusing me with Alexandre Dumas, or maybe Barbara Cartland (Chapter 2). Believe me, plausibility might be hard work, but flawed is dead easy.

Signalling by the topic sentence can alert your reader of organization of material to come, creating an expectation that the rest of the paragraph fulfills:

Carbon-oxygen double bonds occur in ketones, carboxylic acids, and esters, but these classes differ in the groups bonded to the carbonyl carbon. In ketones . . . In carboxylic acids . . . In esters . . .

Alternatively, you can take advantage of reader expectations by adopting a standard organizational scheme. There are a number of possibilities, each of which matches a natural way that readers tend to organize their own thinking. In addition, these are so commonly used by writers that readers are practiced at following them. You might use organization that is:

- **Spatial or temporal**. This works particularly well for description: for instance, of a stratigraphic section from bottom to top, or an experimental procedure from beginning to end.
- **General to specific**. This scheme introduces a general topic and then narrows it to finer and finer detail. This is particularly effective in an Introduction (chapter 10).
- **Specific to general**. This scheme reverses the previous one, building from details to a broader conclusion. This scheme is particularly effective in a Discussion (chapter 13).
- **Least to most important**. This emphasizes the building of an argument toward the most important point in the power position at the end of the paragraph.
- **Familiar to unfamiliar**. This scheme accommodates readers' preferences for beginning on comfortable ground, then extending themselves to the unfamiliar.
- **Simple to complex**. Along similar lines, readers prefer to deal with simple items before more complicated ones. This suggests placing rules before exceptions and simple special cases before complex models.
- **Certain to uncertain**. Contestable claims are more easily understood, and will be more fairly considered, when they are built on foundations of more settled material. You can take advantage of this by establishing conventional wisdom before showing its limitations or providing a compet-

ing new model. The forces-in-protostar-accretion paragraph above is organized this way.

In pursuit of coherence, good organization is complemented by "relational devices": words or structures that indicate relationships between sentences. (We'll see later that we use similar devices to connect paragraphs.) If you don't use relational devices, you ask your reader to do the work of fitting each bit of information into the paragraph's overall framework. That's really a job that belongs to you as the writer. Common relational devices include:

- **Parallel construction and word choice**. The reader can be led comfortably through a sequence of material when it is expressed in a series of grammatically, linguistically, or structurally similar sentences. The most obvious example is the numbering of points in a list: "First, . . . Second, . . . Third, . . ." Less blatantly, a series of sentences can be parallel in phrasing, sentence structure, and word choice:

  ¶**We measured enzyme activity** in presence of inhibitor with an in vitro assay. **We** first **purified the enzyme** in a sucrose gradient. **We** then **added 0.1 μmol of purified enzyme** to each well of a 96-well plate, and **added 0.1 μmol of inhibitor** to half the wells. **We incubated the plates** at 37° for 30 min, and then **added 0.1 or 1 μmol of substrate** to each well. Finally, **we assayed enzyme activity** spectrophotometrically.

  Notice how each sentence takes the same basic structure, and furthermore how words playing similar functions take parallel form: the same part of speech, the same tense, voice, number, and so on.

- **Repetition**. Repeating key words or phrases can tie sentences together and also remind readers of the paragraph's topic:

  ¶**We** measured **enzyme activity** in presence of **inhibitor** with an in vitro **assay. We** first **purified** the **enzyme** in a sucrose gradient. **We** then **added** 0.1 μmol of **purified enzyme** to each **well** of a 96-well **plate**, and **added** 0.1 μmol of **inhibitor** to half the **wells. We** incubated the **plates** at 37° for 30 min, and **then added** 0.1 or 1 μmol of substrate to each **well.** Finally, **we** assayed **enzyme activity** spectrophotometrically.

Parallelism and repetition are complementary techniques; I used both in the enzyme-assay example.

It can be tricky to hit the right degree of parallelism and repetition. Inexperienced writers often overuse them, making passages choppy and stilted. But when this is pointed out, many writers overcorrect, making sure every sentence uses different sentence structure, voice, vocabulary, and so on. A little variety keeps your reader awake, but too much disrupts coherence: it asks readers to master, with each sentence, not just new material but also a new way of encoding that material linguistically. As in many similar aspects of the craft, you can develop a sense of how much repetition (and how much variety) is enough by paying attention to passages you admire from the work of others (chapter 3).

• **Transitional expressions**. Transitional expressions are words or short phrases that explicitly indicate the relationship of one sentence to earlier ones. Transitional expressions may be adjectives, adverbs, conjunctions, prepositions, or phrases that function in one of those ways: for example, *also, although, as an example, because, hence, however, in conclusion, next, on the other hand, similarly, specifically, that is, then,* or *until then*. (Fowler and Aaron [2011], among other sources, offers a much longer list.)

Compare two versions of this paragraph:

> Without transitional expressions: "¶Lava from the 1998 eruption was richer in metals than that from the 1983 or 1977 ones. The 1998 lava was denser and less CO_2-rich. Ejecta volumes were very similar. We cannot reject the hypothesis that eruptions are driven by periodic filling of a single magma chamber."

> With transitional expressions: "¶Lava from the 1998 eruption was richer in metals than that from the 1983 or 1977 ones. The 1998 lava was, **furthermore,** denser and less CO_2-rich. **Despite these differences**, ejecta volumes were very similar. **As a result**, we cannot reject the hypothesis that eruptions are driven by periodic filling of a single magma chamber."

The two versions contain exactly the same information, but the second is much easier to understand because the transitional expressions draw strong connections between different bits of information.

Making paragraphs distinct. Paragraphs should be logically distinct: each with its own topic, and each providing a complete and self-contained treatment of that topic. The meaning of "complete and self-contained" is a bit subtle, though. Each paragraph you write should announce a new topic (or a new angle on a larger topic) and then bring its development to some conclusion. But this certainly doesn't mean paragraphs are independent, as is frequently claimed. If you're writing about something even the slightest bit complicated, your readers' understanding of each paragraph will depend on their assimilation of earlier ones. Not only that, but paragraphs need to work together in building your manuscript's larger argument. So a new paragraph should neither represent a whole new train of thought nor just continue its predecessor's. It should take the train onto a new track—but a new track that's part of the same overall journey.

For paragraphs to work together despite their distinctness, readers need to understand how each relates to the one before. Techniques for linking paragraphs are essentially the same as those for linking sentences within paragraphs: logical arrangement of paragraphs within sections, parallelism in structure and wording, repetition, and transitional expressions. The tricky part is that these linkages should be strong enough to ease the reader's flow through the argument, but not so strong that distinctness is compromised. Linkages that are too strong are common. This problem usually take one of three forms:

- **Superfluous paragraph breaks**. Sometimes a writer will feel a paragraph is getting too long and try to fix the problem by introducing a paragraph break—without actually engineering a shift in topic. This mistake is easy to recognize, because the content of the second paragraph will belong to the topic declared by the first. If you want to divide a paragraph, you need to clearly divide its topic.

- **Overstrong dependencies**. A paragraph may start with a pronoun or transitional expression that doesn't just *link* it to the preceding paragraph, but makes the meaning *entirely dependent on* it. For example, a paragraph might open "¶**Despite this,** it remains clear that . . . ," or "¶**For example,** Devonian trilobites . . ." Either of these constructions would nicely connect sentences within a paragraph, but at the beginning of a paragraph each demands reference to information that is not there.

The solution to overstrong dependencies is to name both halves of the linkage in the linking sentence. You can replace a pronoun with repetition of its antecedent: "¶Despite the difference between replicate experiments, it remains clear that . . ."), or add a brief reminder of the preceding paragraph to fill out the transitional expression: "¶Devonian trilobites provide a good example of a declining lineage . . ."

- **Redundant paragraphs.** A section may include two paragraphs (adjacent or not) that treat substantially the same topic. This adds unnecessary length (chapter 21) and also creates confusion: readers who expect distinctness will wonder what they are missing.

Organizing Paragraphs into Sections

With the assembly of sentences into paragraphs under our belts, we can zoom out just a little to revisit the assembly of paragraphs into sections. In chapter 7, I pointed out that the topic sentences of each paragraph constitute a kind of outline that you can use to plan the telling of your paper's story. There I largely ducked the issue of how to arrange these points to make a smoothly building argument. Now I can be a bit more explicit: all the techniques that build sentences into coherent paragraphs can also help build paragraphs into coherent sections. Sections flow well if paragraphs are arranged spatially or temporally, from least to most important, from simplest to most complex, and so on. Parallelism among paragraphs in structure or repetition of wording favors overall coherence too.

An especially important form of parallelism in scientific writing is consistency in paragraph order from section to section. Often, you will include analogous sets of paragraphs in Methods and Results, in Results and Discussion, or in Introduction and Discussion. Arranging paragraphs in the same order in each section allows your later sections to take advantage of reader expectations established by the earlier ones.

While planning the organization of paragraphs into sections is one major goal of outlining, don't forget that it's quite normal for your story to change as you develop the outline into a manuscript (chapter 5). Expect, therefore, to revisit the order and organization of paragraphs as you write and revise.

Chapter Summary

- A paragraph introduces and treats a single idea.
- Good paragraphs are *unified*, *coherent*, and *distinct*.
- A paragraph is unified when all content relates to the idea declared in its topic sentence.
- A paragraph is coherent when it develops an idea in a logical order with each step related to the next.
- A paragraph is distinct when its treatment of an idea is complete and self-contained.
- Just as sentences are assembled into paragraphs, paragraphs are assembled into sections; the same guidelines apply.

Exercises

1. Choose a recently published paper in your field, and from it choose two consecutive paragraphs in either Introduction or Discussion.
    a. What is the central idea of each paragraph?
    b. Are the two paragraphs distinct? How?
    c. Are they connected by relational devices?
2. Now focus on a single longish paragraph (at least eight sentences).
    a. How is material organized within the paragraph (e.g., spatially, temporally, general-to-specific, least to most important)?
    b. Rewrite the paragraph twice, with the same content, but each time with a different organizational scheme. Which of the three versions (including original) is most effective? Why?
    c. In the original, identify some relational devices that lead the reader through the paragraph. In your rewritten versions, what relational devices did you use?

EIGHTEEN

||||||||||||||||||||||||||||||||||||||||||||||||||||||||||||||||||||||||||||||||||||||||||||||||||||||||||||||||||||||||||||||||||||||||||||||||||||||||||||||||||||||||||||||

Sentences

My discussion of paragraphs (chapter 17) presumed a basic familiarity with their building blocks: sentences. This wasn't too desperate a leap, as by the age of two or three, most children can produce simple but recognizably structured sentences. *Mastery* of sentence construction is another matter. English provides a bewildering variety of ways to build more and more complex sentences, which is why one standard composition book (Fowler and Aaron 2011) allots 259 pages to sentence construction. Most scientific writers experience at least occasional difficulties. Partly this is for the same reason I emphasized in considering paragraph structure: our subject matter is complex. In addition, some issues (involving tense and voice, for instance) are specific to scientific writing.

Grammar

Tackling sentence construction means confronting something that inspires in many scientific writers either boredom or loathing: grammar.

Those of us with English as a first language often remember being hectored mercilessly in high school about the rules of English grammar. "Never end a sentence with a preposition," we were told, "and never start a sentence with 'And.'" Writers newer to English (see chapter 27) usually discover similar advice in courses, books, and online guides. The thing is, while this advice (and a lot more like it) is nearly universal, it's also wrong. It's wrong in detail—sometimes "And" is just the right way to start a sentence—but more important, it's wrong in concept. If you see grammar as a long list of rules for distinguishing "correct" sentences from "incorrect" ones, you risk losing sight of its real purpose: to help you achieve the crystal clarity that makes your writing seem telepathic. Think of grammar instead as a set of agreed-upon conventions by which your reader can easily assign to a sequence of words the meaning that you intended. Good writers respect these conventions because readers easily understand sentences that adhere to them[1]. This is the only reason—but an excellent one—to take grammar seriously.

Knowing the rules of grammar needn't mean slavish obedience to them. In fact, occasional violations of the formal rules can make your writing more effective. Only a writer who knows the rules well, though, can distinguish between three classes of rule violations: those that disrupt meaning, those that merely sound unprofessional, and those that enhance the writing.

- **Violations that disrupt meaning**. The most serious grammatical errors lead readers to infer something different from what the writer intended. For instance, imagine that you read this in a Methods section:

    > After thorough washing with Eliminase, epibionts were scraped from the kelp and processed for DNA amplification and sequencing.

    You are also interested in organisms growing on the surfaces of kelp ("epibionts"), so you collect your own kelp, thoroughly wash it and its epibionts with Eliminase, scrape them off and amplify DNA for sequencing—and get nothing. After months of fruitless work and thousands of wasted dollars, you contact the author, who replies with surprise that you should have washed *the scraper* with Eliminase, not the kelp. Now you understand: Eliminase destroys DNA, and would have prevented contamination from the scraper. Instead, in washing the kelp, you destroyed the very DNA you wanted to sequence.

    This costly grammatical error was a dangling modifier: "*after thorough washing*" was meant to describe treatment of the scraper, but

---

[1] As an example: English readers expect sentences to be ordered subject-verb-object. This is arbitrary, and other languages use other orders, but to an English reader other orders sound amateurish and can even be misleading. To see how violating conventions can produce awkward writing, tune your radio to a classic-rock station and wait the ten minutes or so it takes for Rod Stewart's "Maggie May" to come on. This clunker of a couplet violates the sentence-order convention: "I laughed at all of your jokes/My love you didn't need to coax" (subject-verb-object/object-subject-verb). It rhymes, but that's about it.

taken at face value it actually describes treatment of the epibionts. You assigned the meaning the grammar told you to, but not the one the writer intended.

- **Violations that sound unprofessional**. Much more common are grammatical errors that don't imperil meaning, but give the reader the impression that you wrote carelessly: "Image post-processing was required to resolve the planet from it's rings." That should, of course, be "its rings." Or consider this dangling modifier: "One rat died prior to taking final blood-pressure measurements . . ." Readers won't actually believe that rats take blood-pressure measurements (the grammatical meaning), but they will cheerfully believe that the writer can't construct a coherent sentence.

  Because errors like these detract from the professional tone of your writing, they raise doubt in the reader's mind about your authority as a writer or even as a scientist. An irritated reader might even set your work aside in favor of papers that are more polished and easier to read.

- **Violations that enhance writing**. The best writers occasionally, and deliberately, break the rules of grammar. You should break them too, but only when you can explain to yourself (1) what rule you're breaking, and (2) why breaking it will make your writing clearer and more compelling.

  As an example, consider the split infinitive. The English rule is that the infinitive form of a verb (*to write, to wash*) shouldn't be interrupted by modifying words ("it is important *to wash* the scraper thoroughly," not "it is important *to* thoroughly *wash* the scraper"). This rule doesn't actually make much sense (Johnson 1991), but nevertheless, readers expect writers to follow it and its frequent violation gives your writing an odor of sloppiness. Sometimes, however, splitting the infinitive conveys meaning more clearly or makes your writing more engaging. That's why the mission of Captain Kirk and the Enterprise is "*to* boldly *go* where no man has gone before": the split infinitive emphasizes the boldness, rather than the going, and the technically correct alternative "*to go* boldly where no man has gone before" falls flat. (It says something unfortunate that more ink has been spilled objecting to the perfectly reasonable split infinitive than to the noninclusive language.) Most rules lend themselves to similar deliberate violation—as long as the violations are occasional, and each is clearly justifiable.

Simple Sentences

The simplest sentences have just two components: a subject and a verb ("Ophelia died"). It's only a minor complication to add a third component: an object to receive the verb's action ("We stirred *the solution*"). Most writers form such simple sentences correctly and easily, of course, but it's worth paying attention to these basics because even the most complex sentences are built by elaborating upon the simple subject-verb-object motif. Not only that, but already we have enough about sentence construction to tackle two issues of special interest to the scientific writer: *tense* and *voice*.

Tense. A verb's tense places its action in past, present, or future. This seems simple, but there are three complications. First, English has twelve tenses, not just three. Second, choices among tenses can be a bit tricky, shifting among and within sections of a scientific paper. Third, in citations tense carries some subtle implications.

Native English speakers may be surprised to learn that they are using twelve tenses. (Those with English as an additional language have probably been taught them explicitly.) For each time (past/present/future), there are four tenses. The first is for actions occurring in the moment or for which duration is not important: *I wrote, I write, I will write* (simple past, present, and future). The second is for actions ongoing through time: *I was writing, I am writing, I will be writing* (past, present, and future progressive). The third is for completed actions: *I had written, I have written, I will have written* (past, present, and future perfect). Finally, the fourth combines the "passage of time" and "completed" implications: *I had been writing, I have been writing, I will have been writing* (past, present, and future perfect progressive). Phew!

The major challenge in scientific writing is to choose the right tenses for your meaning, while avoiding frequent tense shifts (which can be jarring for the reader). Tense usage tends to differ between sections. To start with the most obvious, your methods will usually be described in the simple past: "We *dried* collected goldenrods at 70ºC and *measured* their dry masses." Other past tenses appear only to serve particular functions. For instance, you might write "we measured dry mass after plants *had dried* for at least 48 hours" (past perfect) to emphasize the completion of drying (which matters because retention of some residual water could distort the data). Present and future appear rarely in Methods sections, with one exception: methods *papers* (which have the primary purpose of offering protocols for others to use; e.g., Idzik et al. 2014) sometimes use the present.

Your Results section, too, will mostly use the simple past: "Goldenrods under herbivore attack *had* greater dry mass than unattacked plants." However, most writers use the present tense to report statistical analyses or to refer to figures and tables: "This difference *is* significant," or "Figure 2 *compares* impacts of gallmaking and leaf-chewing herbivores." This makes sense from the temporal perspective of the reader: at the moment of reading, the difference is still significant and the figure still compares impacts, even though the measurements were taken in the past.

Tense usage will be more variable in your Introduction and Discussion. The Introduction often uses the past perfect (to indicate study over time) and simple present (to indicate current understanding): "Herbivore preferences *have been* studied for decades, and yet it *remains* unclear whether plant size or nutritional quality influences preference most." Payoff to the work may be suggested using the simple future: "Detailed understanding of herbivore preferences *will allow* prediction of plant-herbivore coevolution." The Discussion may do any of these things, plus offer interpretation of your results. This involves mixing past and present tenses: "That attacked plants *had* greater dry mass *suggests* that herbivores prefer more vigorous hosts." The data are past, but the interpretation is ongoing.

Finally, citations can use simple past, present perfect, or simple present (at least). These tenses imply slightly different attitudes toward the cited article:

- "†Ng (2006) *reported* preference by a caterpillar for phosphorus-enriched leaves." The simple past emphasizes the (past) act of reporting rather than the finding. This may be appropriate if the passage traces historical understanding, if phosphorus preferences haven't been widely documented, or if there is reason to doubt the report.

- "Several reports *have documented* herbivore preferences for phosphorus-enriched leaves (e.g., †Ng 2006, †Muñoz 2012)." The present perfect emphasizes a continuity of reporting and suggests greater generality of, or credence in, phosphorus preferences.
- "Herbivores prefer phosphorus-enriched plant tissues (e.g., †Ng 2006)." The simple present treats phosphorus preferences as established scientific knowledge, with the particular citation receding into the background.

Voice. If any point of grammar can get a roomful of scientists spitting mad, it's the matter of active vs. passive voice (active *I felled ten trees* vs. passive *Ten trees were felled*). Notice the differences. In the active voice, the subject of the sentence is a scientist performing the action, and the object is "trees" (well, technically, the noun phrase "ten trees"). In the passive, the subject "trees" receives the action, there is no object, and the scientist is not mentioned. These features make a dramatic difference to the tone of your writing:

Passive: "Six sites were selected in spruce-hemlock forest and two 50 x 50 m plots were marked in each site. One plot of each pair was chosen at random, and the four mature spruce trees closest to its center were marked and then wounded by being struck 10 times with a hatchet. Four months later, all marked trees were felled. A 1.5 m bolt was cut from each felled tree, and each bolt was returned to the laboratory and held in a mesh cage at 15°C for collection of emerging insects."

Active: "I selected six sites in spruce-hemlock forest and marked two 50 x 50 m plots in each site. I chose one plot of each pair, marked the four mature spruce trees closest to its center, and wounded each tree by striking it 10 times with a hatchet. Four months later, I felled all marked trees. I cut a 1.5 m bolt from each felled tree, returned the bolts to the laboratory, and held them in mesh cages at 15°C to collect emerging insects."

Legions of undergraduates have been told that scientists should write in the passive voice (and never, ever, write "I"). This advice is wrong. The passive *is* prevalent in the literature—but it hasn't always been, and the tide is shifting back toward the active.

Early scientific writing was predominantly active-voice (Gross et al. 2002). This fit well with science done by respected gentlemen and with authority derived from virtual witnessing (Box 11.1): vivid description of the actors and the action conferred rhetorical strength. As science became professionalized in the nineteenth century, however, scientists looked for objectivity in prose—with objectivity meaning "knowledge that bears no trace of the knower" (Daston and Gallison 2007). The passive voice let writers suppress any mention of the person who actually conducted the experiment, analyzed the data, or drew the conclusion. This is odd, though, because we all know it's only pretense: trees don't fell themselves! Authority in modern science comes from our adoption of appropriate conduct and techniques—not from pretending we don't exist.

With implications for authority removed, what case remains for routinely using the passive voice? The best I can do is that the passive matches the grammatical subject of the sentence to the logical subject of the writing. That is, "I felled ten trees" emphasizes the scientist, while "ten trees were felled" emphasizes the trees. It's really the trees we're writing about, and the passive voice puts them up front. This advantage is very much outweighed, though, by the multiple and important advantages of the active voice. For at least five reasons, the active voice helps you achieve clear communication with your reader:

- **It's shorter.** My tree-felling paragraph is seven words (eight percent) shorter in the active voice. Chances to reduce length without cutting content are precious (chapter 20).
- **It's easier to read.** The active voice is more common in everyday reading material. On top of that, passive constructions are often convoluted. For both reasons, passive-voice writing asks more effort of the reader.
- **It's more engaging.** The active voice recognizes the human actor. Readers relate more naturally to passages about people, with whom they identify, than to passages about objects or abstractions.
- **It's more vivid.** The passive voice forces the use of vague and dreary verbs (especially, endlessly repeated forms of *to be*). The active instead uses forceful, action-packed verbs (*chose, wounded, felled*).
- **It's more honest.** The active voice acknowledges that a person (the writer) did the work. Why pretend otherwise?

The argument for the active voice is powerful, but I wouldn't suggest stamping out every trace of the passive. Varying voice a bit can make a passage less monotonous. Furthermore, the passive voice is well suited to a few specific functions:

- **Directing attention.** When the active voice predominates, a contrasting use of the passive can call special attention to something by moving it into the subject position ("Beetles were identified to species, but other taxa only to genus").

- **Obscuring the actor.** Sometimes we don't know or care who or what was responsible for the verb's action, and we can use the passive to move the actor into the background: "Non-parametric methods are preferred when data violate regression assumptions." Since this preference is broadly accepted, the active "Statisticians prefer non-parametric methods . . ." adds nothing.

- **Avoiding a complex subject.** Sometimes the active voice would have a complicated subject—perhaps some complex entity that the reader has not met before. For instance, I might want to describe the insects that attacked my spruce trees: "*Buprestid and cerambycid beetles, woodwasps, and several other taxa that damage trees or that vector pathogenic fungi* attacked the wounded trees." This is unwieldy, though. A passive version is much clearer: "*The wounded trees* were attacked by buprestid and cerambycid beetles, woodwasps, and several other taxa that damage trees or that vector pathogenic fungi." This gives the reader a simple, familiar *subject* and proceeds quickly to the verb. Only with that core of the sentence established is the reader asked to consider unfamiliar content.

In summary: use the active voice unless you are sure that the passive makes a particular sentence more effective.

How Sentences Become Complex

Sentences in scientific writing are rarely as straightforward as "I felled ten trees." Instead, writers add complexity in an enormous variety of ways, as sentences become elaborated with modifiers, phrases, and multiple clauses. For example, I might write "We marked healthy, mature

white spruce trees 15 to 20 cm in diameter at breast height with individually numbered aluminum disk tags fastened to the lowest available
branch with nylon cable ties." Noun strings replace single nouns (e.g.,
"aluminum disk tags," with the nouns "aluminum" and "disk" modifying
"tags"). Long phrases loaded with modifiers replace either nouns (e.g.,
"healthy, mature white spruce trees 15 to 20 cm in diameter at breast
height," acting as the object of the verb "marked") or adjectives (e.g.,
"fastened to the lowest available branch with nylon cable ties," modifying "aluminum disk tags"). All this is elaborate, but the sentence can still
be boiled down to a single clause with one subject, verb, and object
("We . . . marked . . . trees"). Further complexity arises when sentences
include multiple clauses, each with its own subject-verb-object structure. Clauses can be independent or subordinate: an independent clause
makes a complete sentence if excised to stand on its own, while a subordinate clause does not. For instance, this sentence includes one **subordinate** and two *independent* clauses: "**Because we kept temperatures constant**, *insects lacked seasonal cues*, and *emergence may have been delayed.*"
There is no limit to the number of clauses a sentence can contain, and
the potential for mayhem is obvious.

While our sophisticated knowledge of the natural world requires that
our writing use some complexity, this doesn't mean it has to be hard to
read. It does mean that we should use strategies that help keep complex
sentences effective. Here are three: pay attention to each sentence's simple core, design sentence structure to work with readers' expectations
and cognitive biases, and limit the degree of complexity.

Each Complex Sentence Has a Simple Core

When a complex sentence gives trouble, it helps to drill down to its simple subject-verb-object core, and then mentally add back complicating
modifiers, phrases, and clauses one by one. (Try this now with the
spruce-marking sentence in the last section.) This works because complex sentences must follow the same basic rules as simple ones. For example, errors of subject-verb agreement often sneak into complex sentences, as in "Research directed at understanding influences of tree
stresses on attack rates by wood-boring insects are extremely impor-

tant." Stripping this sentence down to its core, "Research . . . are . . . important," makes the error obvious: the verb should agree with its subject ("research"), not the nearest noun ("insects"). Errors like this happen when you're distracted by thickets of modifiers and complex structure. Trim away the thickets, fix the underlying structure, and then—only to the extent necessary—carefully put the thickets back.

Readers Have Expectations and Cognitive Biases for Sentence Structure

Even a complex sentence can be crystal clear if it's built to exploit the way your reader thinks (Gopen and Swan 1990). Readers have cognitive biases (built-in ways of processing material during reading) and expectations (learned anticipation that text will adhere to common structural conventions). Working with these biases and expectations lets you present information just as the reader naturally looks to receive it. This perspective leads to two important principles: well-written sentences keep actions close to actors, and they sort material into natural topic and stress positions.

Actions and actors. Readers are hardwired to look for actions and actors (verbs and subjects) and apply one to the other, but this is challenging when these elements are too far separated. Consider this sentence:

> Three results—that wounded trees were attacked by more insects, that this pattern was stronger at sparsely forested sites, and that more insects emerged from trees on the downwind side of the plots—suggest that females use plant volatiles to find hosts.

The subject/verb pair here is "results suggest," but three clauses and thirty-one words interrupt it, so the reader is left hanging for most of the sentence's length. Revising to unite actor and action fixes the problem:

> Three results suggest that females use plant volatiles to find hosts: wounded trees were attacked by more insects, this pattern was stronger at sparsely-forested sites, and more insects emerged from trees on the downwind side of the plots.

Topic and stress positions. Just like the first and last sentences of paragraphs (chapter 17), the beginning and end of each sentence represent power positions. You will write most clearly if you realize that readers expect to find particular kinds of information in each position. The opening few words of a sentence are the *topic position*, which readers automatically take as an orientation to what the sentence is about. The end of the sentence is the *stress position*, and readers ascribe importance to whatever they find there. (In complex sentences readers tend to assign topic and stress positions within each clause, although with multiple stress positions in a sentence the impact of each one becomes diluted.) An easily-read sentence starts readers in comfortable territory by placing familiar, or at least simple, information in the topic position. Then it builds toward new, more complex, information in the stress position.

Ignoring reader expectations for topic and stress positions impedes clear communication. For example:

> We used high-performance liquid chromatography to analyze reaction products. We used a Kinetex XB-C18 reverse-phase column for separation and UV absorbance to detect separated compounds. A 10% to 90% water/acetonitrile blend (20 min linear gradient) was the mobile phase. We added 10 uL of 0.1 mM phenylbutazone to each sample to serve as a run-time standard.

This paragraph makes the reader work too hard. Each sentence begins with new information, and only at the end reveals what that information is for. This forces the reader to store information in short-term memory until it can be related to the overall logic ("they added phenylbutazone—I wonder why—better remember that until I find out"). Furthermore, the reader will ascribe importance to information in the stress positions—but it's the wrong information! For example, the first sentence stresses the reaction products, but the paragraph is really about the analysis; and the third sentence stresses existence of a mobile phase (obvious to any reader familiar with HPLC) rather than its nature.

Careful thought about *topic* and **stress** suggests this revision:

> *We analyzed reaction products* using **high-performance liquid chromatography**. *Separation* was performed on a Kinetex XB-C18

reverse-phase column, and *detection* was by **UV absorbance**. *The mobile phase* was a water/acetonitrile blend with a 20 min linear **gradient from 10% to 90% acetonitrile**. *To provide a run-time standard*, we spiked each sample with 10 uL of 0.1 mM **phenylbutazone**.

Now each sentence places familiar material in its topic position, and then develops that topic to add new and important information. Sometimes the topic material is familiar because it has appeared previously; for instance, in the second sentence chromatographic separation is recalled from the stress position of the first. (This stress-to-topic evolution can draw your reader smoothly from sentence to sentence; most readers find the first version choppy.) Other times, the topic material references readers' prior knowledge: this passage is written for readers familiar with HPLC, who know that there must be a mobile phase (solvent flowing through the column) but need to be told what solvent was used. In each sentence the stress position introduces new material, which becomes the takeaway message for the reader: the use of HPLC, a reverse-phase column, UV absorbance, an acetonitrile gradient, and a phenylbutazone standard.

Limiting Complexity

Writing about science requires *some* complex sentences, but that's not a license to go crazy. It's not hard to find examples of very complex sentences in the scientific literature, and some writers seem convinced that complexity brings authority. Remember, though, the value of clarity in light of all the competing demands on your reader's time and energy. Use the simplest sentences that can accommodate the material you need to explain, and interrupt blocks of long, complex sentences with short and simple ones. (The variety in cadence makes for livelier reading.) Online tools (e.g., http://www.hemingwayapp.com) can help you spot some, but of course not all, complex-sentence problems.

It isn't possible to catalogue all the ways that sentences grow too complex, but two problems seem especially common in scientific writing. The first is the use of long, complex modifiers. Among the worst are noun strings: phrases built of consecutive nouns, each modifying the next and working together as a single subject or object (like the five-noun string in "We developed a *gibbon leucocyte cytoplasm fractionation protocol*"). Strings longer than two or three nouns are difficult for readers to parse, and should be broken up even if that requires a few extra words ("We developed a protocol for fractionating cytoplasm from gibbon leucocytes").

The second common problem is the sentence that takes its reader on a long and roundabout journey. This is not the same thing as a long sentence: a well-crafted sentence can be long and still build smoothly to its point. But the more material appears in a single sentence, the harder it is to craft, and the more likely that the reader will feel taken here, there, and everywhere. In large part this is because including more material makes your decision about what to place in topic and stress positions simultaneously more important and more difficult. I once read this sentence in a thesis:

> Using data obtained from our genetic structure analyses, we selected markers that showed polymorphism in at least 50% of the populations to examine allele frequencies along environmental gradients to determine whether environment influenced genetic composition.

This is a very complex sentence, introducing a lot of information via at least nine "chunks" (phrases and clauses):

> [1]Using data [2]obtained from our genetic structure analyses, [3]we selected markers [4]that showed polymorphism [5]in at least 50% of the populations [6]to examine allele frequencies [7]along environmental gradients [8]to determine whether [9]environment influenced genetic composition.

Which chunk(s) should be the topic material, which the stressed, and which should connect topic to stress? I don't know. Chunk nine ("environment predicted genetic composition") occupies the stress position, but I think it's really the topic. It isn't clear what should be stressed instead. Placement of the topic at the end is a common mistake, and results in what I call a "stack-loading" sentence. To understand the genetic-composition sentence, you need to load eight chunks successively into short-term memory, retain all of them until you reach the last chunk and discover the underlying topic, and then recall them one by one to add their content to your understanding. This is how computer programs load and unload data to and from stacks during complex calculations—but computers are very good at handling stacks, while humans are not. A confused reader has two choices: read the sentence over and over, or proceed without understanding it.

So our genetic-composition sentence has two related problems: too much content, and unhelpful arrangement. Here's one way we might rescue it (indicating *topic* and **stress**):

> [8/9]*If environment influences genetic composition,* [6]allele frequencies should change [7]**along environmental gradients**. [*]*We tested this hypothesis* [1]using data [2]**from our genetic structure analyses,** [3]selecting markers [4]**that showed polymorphism** [5]**in at least 50% of studied populations**.

There are now two sentences, the second with two stress positions. All the same content appears, plus a new chunk (*) that recapitulates the stressed material of the first sentence as the topic of the second. No stack-loading is required to understand the revision: each sentence declares its topic and then develops it in a straight line into new, stressed material. Pinker (2014, his chapter 4) calls this a "right-branching" sentence, and discusses sentence structure at length.

To summarize: you can write complex sentences and still communicate clearly, if you construct them with care. However, if you go astray you run the risk of complexity misleading or repelling your reader. In general, therefore, view complex sentences with healthy suspicion, and simplify when you can.

Chapter Summary

- The rules of English grammar are important, but can sometimes be violated in ways that make writing more effective.
- A simple sentence has a subject, a verb, and usually an object.
- Scientific writing uses a variety of verb tenses. Methods and Results sections use mostly the simple past, but Introduction and Discussion sections can mix tenses to distinguish past study from current understanding and future prospects, and to indicate attitude toward cited papers.

- You should write in the active voice as much as possible, despite the scientific literature's fondness for the passive.
- Complex sentences are inevitable, but you can minimize their load on readers by taking advantage of reader expectations for sentence structure—for instance, proximity of actions and actors, and placement of important material in topic and stress positions. Reducing complexity when you can is a good idea.

Exercises

1. Choose a recently published paper in your field, and select a passage from the Methods (two to three paragraphs, at least five hundred words). If the passage is written in passive voice, rewrite it in the active; or vice versa. Which version is longer? Which do you find easier to read? Why?

2. Choose a long sentence from the Methods section you're working with (at least thirty-five words).

   a. Underline subordinate and independent clauses.

   b. Rewrite the long sentence as series of short ones (no more than ten words per sentence). Is the result clearer or less clear? Why?

   c. Identify each subject/verb pair. How many words separate each pair? For the farthest pair, rewrite the sentence to bring them closer together.

   d. What appears in the topic and stress positions of the sentence? Why did the writer make those choices?

NINETEEN

||||||||||||||||||||||||||||||||||||||||||||||||||||||||||||||||||||||||||||||||||||||||||||||||||||||||||||||||||||||||||||||||||||||||||||||||||||||||||||||||||||||||||||||||||

Words

The English language includes at least 350,000 words, maybe more than a million[1], and is growing rapidly (Michel et al. 2011). This extraordinarily rich lexicon means that you'll often face the challenge of choosing the best word among several, or even many, options. For example, any of the four words in parentheses could complete this sentence about sugar-polymer content in soybeans:

> Soybean cultivar EG7 had higher (complex carbohydrate/polysaccharide/starch/amylose) content than the other studied cultivars.

The best choice is the one that makes your writing clearest—but which one is that, and why? Consider these three factors: what words mean, what the writer wants to say, and who the readers are.

What words mean. It's obvious that you should know the meaning of each word you use, but that's not as simple as it sounds. A word's "meaning" is a convention shared by a set of readers about the information conveyed by its use. That is, what *you* think a word means really isn't relevant; instead, you need to know what meaning your readers will assign to it. That would be easy if English were like Morse code, with a universal and unchanging 1:1 mapping of symbol to meaning. (All agree that dot-dot-dot, and only dot-dot-dot, means *S*.) But English is much more interesting than that, with potential for ambiguous meanings or

---

[1] It isn't really possible to count. Are *pencil* and *pencils* two different words? *Cow* (the animal) and *cow* (to intimidate)? The *cowl* worn by a monk and the *cowl* around an aircraft engine? Do proper nouns count? Chemical names? Archaic words? Foreign borrowings (longstanding ones like Arabic *algebra* or Algonquian *raccoon*, or new ones like Japanese *anime* and *edamame*)? Before vigorous exercise, one should warm up; so before debating the number of English words one should probably settle a few easy arguments—perhaps about the designated-hitter rule or the merits of affirmative action.

for disagreement among readers. English has multiple words for almost every concept, some completely synonymous, some differing slightly in connotation, and some overlapping but with important distinctions. Single words often have multiple meanings, some more familiar than others. English even has words that are their own antonyms (*cleave* can mean to stick together or to cut apart) and pairs of words that are both antonyms and synonyms (*best* and *worst* are antonyms as adjectives, but synonyms as verbs—as in to *best/worst* someone in a fight).

As a further complication, meanings can evolve and sometimes even reverse over time: *counterfeit*, for instance, originally meant a genuine copy. *Counterfeit*'s meaning has stabilized, and it would confuse a reader now only if used in its original sense; but other words are currently in states of rapid semantic drift, and are best avoided. Consider, for example, what happens if you mention a *moot* point. Some readers will recognize the original meaning (a point open to debate), others will take the new sense (a point irrelevant to the matter at hand), and still others won't be sure which meaning you intended.

Fortunately, for most words, the social convention defining meaning is relatively stable and is shared by most readers. A standard dictionary provides a useful record of these meanings, and you shouldn't be ashamed to use one. Words for which convention is looser (those with many shades of meaning, or with meaning currently mid-drift) are words you may want to avoid.

What the writer wants to say. Meaning alone doesn't determine the right word to use. That's because words with different meanings offer you choices about exactly what information you communicate. Such choices go back to your decisions about the story you want to tell (chapter 7) and the information your reader needs in order to understand that story. Take our soybean-cultivar example: "Soybean cultivar EG7 had higher (complex carbohydrate/polysaccharide/starch/amylose) content than the other studied cultivars." It might well be that cultivar EG7 is higher than the other cultivars in *complex carbohydrates*, in *polysaccharides*, in *starch*, **and** in *amylose*. If so, then any of the four word choices makes for an accurate statement—but which statement fits your story best? *Complex carbohydrate* and *polysaccharide* both refer to any polymer of sugar, while *starch* and *amylose* are both polysaccharides with α-bonded glucose units but starch includes both straight-chain amylose and its branched cousin amylopectin. If your story is about calories in diet, then *starch* is the word you want: *complex carbohydrate* and *polysaccharide* include indigestible fiber, while *amylose* excludes calorie-containing amylopectin. If your story is about the plant's carbon allocation, you might choose *polysaccharide*; or if it's about biosynthetic enzymes, *amylose*. In each case one choice best matches the story you've decided to tell (and nearly every word you write offers similar choices). Of course, I'm not recommending word choice that misleads: if your assay measured only amylose, then *amylose* must be your choice. You should tailor word choice to your story, but never steer the reader toward anything other than a clear understanding of what is true.

Who the readers are. While the meaning of a word is a convention shared among readers, not every reader is a party to every convention. As a result, the right word choice depends on who your readers are and what they bring to your effort to communicate with them. Most readers of your primary-literature journal papers know your field the way you do, so you can assume they're familiar with technical terms (*amylose*), and their assignment of meanings to words won't surprise you much. Readers of review papers or grant proposals may work in fields somewhat removed from your own, and your word choice should recognize this. If you're writing for nonscientific readers (perhaps in a report to government or a lay essay), your word choice must accommodate an audience unfamiliar with technical terms and uninterested in their precise meaning. It isn't just that readers either know a word's meaning or don't: some quite common words have different meanings for different audiences. *Significant* is a good example: it means "important" or "large" to lay readers, but "statistically inconsistent with the null hypothesis" to scientific ones. This difference has led to much miscommunication of medical research, as papers reporting *significant* but small effects of potential treatments spark newspaper stories about wonder cures. It's easy to forget that your readers aren't you—but it's your job as a writer to remember.

These three considerations aren't independent. *Who the readers are* influences *what words mean* to them, *what the writer wants to say* depends on *who the readers are*, and so on. If this seems complicated, remember the single, simple goal: crystal-clear communication with your reader. With effort and experience, and attention to the advice of friendly and formal reviewers (chapters 22–23), you can become increasingly adept at choosing the right word.

Long Words, Short Words, and Jargon

New readers of the scientific literature must confront a lexicon that differs in many ways from what they've encountered before. Scientific writers use many long words, technical words, newly coined words, and condensed words (abbreviations, acronyms, etc.). There are good reasons for each: principally that long, technical, and new words allow very precise meaning, while condensed words conserve page space. The tradeoff is that such words demand mental effort from the reader, so the best writing uses as few as possible.

Long words have been tempting writers for nearly a thousand years. Following the Norman conquest of England in 1066, French became the language of that country's government and Church, while peasants used Old English. As French words were assimilated into English, this contrast led to the view that the (generally longer) French-origin words were more serious and professional and the (shorter) Old English words were coarse and common[2]. Use of Latin as the language of scholarship during and after the Renaissance reinforced this belief, and to this day many writers seem convinced that longer words indicate deeper thought. Not so. Except when they bring precision necessary to your story, replace long words with short ones: *utilize* with *use*, *consequently* with *thus*, *approximately* with *about* and so on. Readers will be grateful.

Technical words almost seem to define scientific writing. Granted, our detailed knowledge of the natural world makes it impossible to report new science without a good deal of precise technical vocabulary: *allosteric*, *baryon*, *disconformity*, *fractal*, *metalloid*. In fact, used for an audience to which they are familiar, technical words both save length and enhance clarity. However, they become jargon if they are used when a simpler word suffices, or when forced on readers unfamiliar with them. Even in the primary literature, it's worth avoiding the most technical terms when readers don't need them. For instance, if you separated

---

[2] Noblemen have *flatulence* (French), but commoners *fart* (Old English). It smells the same.

proteins by polyacrylamide gel electrophoresis, your paper might specify *polyacrylamide gel* once in the Methods, but use just *gel* elsewhere: readers need the additional precision of *polyacrylamide* to understand the methods, but they needn't be reminded when discussing results or interpretation. The broader the audience, of course, the harder you should work to minimize technical terms.

The temptation to coin new technical words is hard for scientific writers to resist. Sometimes, of course, it has to be done: if you are describing a new species, mineral, or astronomical object, it needs a name. Otherwise, remember that your new word will impose a double burden on your readers: it will be not just highly technical but also unfamiliar. As cool as it may seem to immortalize yourself in the language, coin only new words that are absolutely indispensable.

Scientific writing is made denser by a profusion of condensed words: abbreviations, contractions, initialisms and acronyms (initialisms, like *DNA*, are pronounced as sequences of letters; acronyms, such as *ANOVA*, are pronounced as if they were words). The use of condensed words in scientific writing increased sixfold over the twentieth century (Gross et al. 2002), and sentences like this are now unremarkable:

> To evaluate the role of extracellular cAMP in sperm capacitation, 10–15 $\times$ 10^6 spermatozoa/mL were incubated in 0.3% BSA sp-TALP at 38.5°C and 5% CO_2 atmosphere for 45 min in the presence of 0.1, 1 or 10 nM cAMP (Osycka-Salut et al. 2014).

Condensed words reduce the physical length of a passage, but only those extremely familiar to readers (*DNA, cAMP, mL*) reduce reader effort. Their frequency owes much to pressure on available journal space, and with luck online publication will loosen the corset stays and let us breathe a bit more freely. In the meantime, journals increasingly require lists of abbreviations, and good writers use them as sparingly as possible. Newly coined abbreviations are doubly difficult for readers and are rarely a good idea.

Nouns, Verbs, and Nominalizations

Scientific writers love "nominalizations": nouns built from verbs (the noun form usually carrying a suffix like "-ance" or "-ation," as in *nomi-*

nalization itself). If you watch for nominalizations, you'll find them everywhere:

We *conducted an analysis* of the data . . .	for	We *analyzed* the data . . .
The adaptive lens *had* superior *performance* . . .	for	The adaptive lens *performed* better . . .
Our *intention is* to . . .	for	We *intend* to . . .
†*Liu (1995) provided a review of* . . .	for	†*Liu (1995) reviewed* . . .

Other common nominalizations in scientific writing include *agreement, calibration, examination, expectation, investigation, preparation, proliferation* and the like.

Nominalizations make writing turgid because they replace lively verbs with passive, bloated nouns, camouflaging the characters and actions in the story you're trying to tell. Readers best understand stories when nouns name the characters and verbs express the actions. When you use nominalizations, your verbs merely mention that action exists (in the examples above, *conducted, had, is, provided*); the nature of the action is hidden in the nominalization (*analyzed, performed, intend, reviewed*) where finding it takes reader effort. To make things worse, nominalizations generally lengthen your text: most are longer than the verbs they're based on, and it takes extra words to connect them to the containing sentence.

You should avoid nominalizations when you can, but expunging every one would go too far. A few are used so often that they give readers no pause at all: for instance, *evolution*. A nominalization can also allow compact reference to something complex or abstract that's explained at more length earlier in the text. I did this with *nominalization* itself at the opening of my last paragraph; it works there as a relational device smoothing the transition between paragraphs (chapter 17).

Troublesome Words

Some words love to make trouble. Many troublesome words come in pairs that differ markedly in meaning or function but only subtly in spelling (*affect/effect, principal/principle*). Other tricky pairs present fine shades of meaning (*alleviate/allay*) or grammatical function (*which/ that*) for writers to trip over. For some words, the trouble is self-reinforcing: frequent misuse by other writers makes errors familiar and likely to escape your notice (as for *criteria* or *data* used as singular). Finally, some troublesome words just have arbitrary or tricky spellings (*accommodate, necessary*).

Published lists of troublesome words can go on for hundreds of pages, but the list of words that trouble any single writer (that is, you) will be much shorter. It's worth compiling your own list and taping it up where you write. Be proactive and browse available lists, either online (search for "troublesome words list") or in book form (Fowler and Burchfield 1996, Bryson 2004; perhaps surprisingly, both books are witty and fun to read). Make note of entries that belong on your own list. The more you can build and check your own trouble list, the more you can spare friendly and formal reviewers (chapters 22, 23) from having to do it for you.

Beware Word-Choice Paralysis

To end this discussion, an important word of caution: it's easy to become paralyzed by word choice. For most writers it's important to keep momentum while working to complete the first draft (chapter 6). If a few moments' thought doesn't produce an obvious word choice, mark the uncertain choice and move on. You'll need to make the decision eventually, but it will probably be easier later on.

Chapter Summary

- Almost any word in anything you write represents a choice among alternative words. The best choice takes into account shades of meaning, precisely what you want to say, and who your readers are.
- Scientific writing inevitably uses technical vocabulary, but long and unfamiliar words are appropriate only when no shorter or more familiar ones will do.

- Abbreviations, contractions, initialisms, and acronyms make writing dense, and unfamiliar ones should be minimized.
- Nominalizations are common in scientific writing, but their overuse makes writing turgid.

Exercises

1. Choose a recently published paper in your field, and select a Methods passage of about 250 words.
   a. Identify a few words that would be jargon to a reader outside your field. Which are necessary for their technical precision, and which could be replaced by shorter or simpler words?
   b. Underline all the condensed words. Which are common knowledge in your field, and which are unfamiliar? Rewrite the passage to avoid the latter. How much length did you add, and do you think that investment is worthwhile?
   c. Underline each nominalization. Rewrite the passage to remove at least one-third of them. Is the passage longer or shorter? Harder or easier to read?

TWENTY

Brevity

Be brief.

Now, that was fun to write, but if advising writers to "be brief" was all it took, you and I could both just skip this chapter. We can't. I've reviewed, formally or informally, somewhere around a thousand manuscripts over my career, and all but a handful should have been shorter. This is why editors nearly always ask writers to condense their manuscripts, and why submissions, conference abstracts, and term papers have length limits. We've all been advised to be brief, but most of us still need help getting there.

There are three major reasons for this call to brevity. First, it helps publishers: journal pages are expensive to produce, and journals are deluged with submissions. Shortening each paper means journals can publish more of them. Second, brevity helps readers: all of us are swamped with written material competing for our attention (chapter 1), and shortening each paper means we can read more of what's out there. Third, brevity helps writers: shorter writing tends to be clearer writing, so your goal of effortless communication can almost always be advanced by shortening your draft.

What exactly do I mean by brevity? Manuscript lengths are most often expressed as word counts, and it's easy to overemphasize those. It's actually better to count characters than words, because replacing long words with short ones is just as helpful to the reader as reducing word count. Second, counts are only a convenient shortcut. What really matters is not how long a sentence or paragraph is, but how quickly your reader can read and understand it. If adding a little extra text actually facilitates reading, go for it; but keep in mind that removing text is nearly always a more important challenge.

Two Ways to Be Brief

There are two ways to achieve brevity in writing: reduce content, or reduce the text used to convey content. Good writers pay attention to both.

Reducing content sounds like a sacrifice, but if you think of it as sticking to your story, it's easier to appreciate its virtues. If a section, a sentence, a detail, a graphic, or a dataset isn't necessary to the story you're telling, take it out (or better still, don't put it in to begin with, which is the point of outlining and related techniques; see chapter 7). This can pay off strikingly. I recently reviewed a manuscript that identified in its Abstract a simple hypothesis and some methods to test it—but spent most of its Methods and Results, and a good chunk of its Discussion, on unnecessary details of the study system and a pair of completely unrelated experiments. Just by sticking to their story, the authors could have cut the manuscript by at least forty percent and set themselves up to have more impact on more readers. Don't miss this kind of easy opportunity.

You don't have to cut content, though, to cut length. It's amazing how much can be cut from most drafts without removing any information. If you've written a two-hundred-word paragraph, ask yourself if you can convey the same content in 190 words, or 180, or 150. Carefully consider each sentence, each phrase, and each word. If it isn't necessary, take it out. If there's a shorter alternative that's equally clear, use it. Look especially for the common text-bulgers addressed in the next section. Be ruthless: every character of your text should be forced to justify its existence.

These two approaches to brevity are best taken at different stages of the writing process. Work at sticking to your story before and as you write: there's no sense laboring to draft content that doesn't belong in your manuscript. In contrast, don't worry too much about economy of expression until you've completed a first draft, lest you lose writing momentum (chapter 6). You can shorten during self-revision (chapter 21).

Common Text-Bulgers

There isn't really any limit to the list of ways that a manuscript can become overlong. However, it's possible to identify some common text-

bulging habits. Put more positively, these are common and easy opportunities to improve brevity.

- **The passive voice**. The active voice is shorter, along with its other
  advantages (chapter 18).
- **Nominalizations**. Turning verbs into nouns makes writing both
  weaker and longer (chapter 19). You can *reduce* length by not writing
  "*achieve a reduction in* length."
- **Long words**. Whether technical or commonplace, a long word belongs only when no shorter one will do (chapter 19). Don't *endeavor*
  to do something you can *try*, or *terminate* something you can *end*.
- **Roundabout phrases**. Many common phrases can be replaced by
  single words that cut more quickly to the point: for example, *owing to
  the fact that* (=*because*), *notwithstanding the fact that* (=*although*), *is
  able to* (=*can*), *the majority of* (=*most*).
- **Tautologous modifiers**. Watch for modifiers already implied by the
  words they modify: *completely finish*, *may potentially*, *ultimate result*,
  blue in color. (Are you worried your reader will think the crystals
  were blue in shape?)
- **Empty modifiers**. Some common modifiers often lack meaning:
  really, *basically*, *actually*, *indeed*, *quite*, *various*. While these words
  can sometimes help a sentence, they're often just verbal tics fossilized
  in ink.
- **Padding**. Sometimes an entire phrase bulks up a sentence without
  adding information: *the fact of the matter is*, *in our opinion*, *needless to
  say*, *it obviously follows that*, *for all intents and purposes*. In speech such
  phrases allow time for the listener to catch up, or for the speaker to
  think of what to say next. In writing, neither function applies. Delete!
- **Hedging**. It's essential to use hedges (chapter 13) to express the precise strength (or limitations) of the claims you're making. Unfortunately, hedges are like potato chips: once you write one, it's hard to
  stop, and they expand your text like your waistline. As an example, I
  read this in a thesis once: "(Observation) could reasonably be assumed to possibly occur by chance." One level of hedging would have
  done—perhaps "The evidence that (observation) was nonrandom
  was modest ($P = 0.06$)." Excessive hedging adds length, and also
  makes it difficult for readers to take your argument seriously. Evalu-

ate each hedge, and keep only those necessary to avoid misleading readers about the strength of your conclusions. Similar logic applies to the use of emphatics, such as *clearly*, *primary*, and *major*.

- **Metadiscourse**. This is writing that's about the writing rather than about the subject of the manuscript (Williams 1990:40). A piece of metadiscourse may refer to the structure and content of the text ("In this section we report . . ."), to the writer's thoughts about or rationale for the writing ("We believe that . . .", or "It has been observed that . . ."), or even to the reader's act of reading ("Consider the following . . ."). Like hedging, metadiscourse can be used effectively: for instance, to provide organizational cues for the reader. But it can also proliferate. Check metadiscourse skeptically, and remove any that only adds length. For instance, "We believe that our results establish that salinity is . . ." says no more than "Salinity is . . ." Other common excesses of metadiscourse include "In this study we" (just say "We . . ."), "The objective of this study was to . . ." ("We sought to . . .", or just "We . . ."), and "It is important to keep in mind that . . ." (omit).
- **Parentheticals**. A parenthetical is a phrase that interrupts an otherwise complete sentence (or a sentence that interrupts an otherwise complete paragraph). As the name suggests, many are enclosed by parentheses, but they can also be set off by commas, by dashes, or not at all. For example, this passage includes three parentheticals:

"Many recovered meteorites are of the nickel-iron type *in part because these are easily recognized as geologically unusual*. However, most meteorites are actually chondrites. Carbonaceous chondrites, *as opposed to the more common ordinary chondrites*, often contain complex organic molecules *such as amino acids*."

Parentheticals can clarify, illustrate, restrict, or otherwise modify meaning. They can also offer extra information that makes the material more interesting or more useful, even if it isn't crucial. However, parentheticals add length, interrupt the reader's flow through the logic of your argument, and distract the reader from the story you're telling. Identify each parenthetical and ask whether it's really worth the demand it places on your reader[1].

---

[1] This book includes quite a few parentheticals (especially footnotes like this one, and how's that for metadiscourse?). In a shorter writing form I'd have excised them ruthlessly. In a book, where

- **Redundancy.** It's amazing how easily redundancy creeps into writing. Check for redundant content throughout your manuscript, but pay special attention to three common trouble spots:
  - Keeping paragraphs to the intended outline. Each paragraph has its own topic, and should stick to it. Once you've covered a topic, don't go back to it in later paragraphs. If you find that you need to, it probably signals a need for revision to your outline.
  - Separating the content of major sections. It's tempting to repeat details of the Methods in the Results section, and to repeat Results in the Discussion. A few cross-references are indeed useful to readers (chapters 12–13), but be sparing. The Introduction and Discussion may have more overlap, because the Introduction sets up questions to be answered in the Discussion, and because both sections need to address things like the context, goals, and significance of the study. The trick is for the Discussion to *refer to* material from the Introduction while *repeating* as little of it as possible.
  - Efficient use of text, tables, and figures. Remember that there's a best way to present any particular dataset or pattern (chapter 12)—but one way will suffice. Don't present the same data in a table and in a figure, for instance. Text referring to a figure or table should provide just enough description that the reader knows what pattern (s)he is being asked to see: for instance, "reaction yield was inversely related to temperature (Figure 1)." More detailed description is wasted.

Getting to Brevity: An Example

To illustrate some of these text-bulgers in a real piece of scientific writing, here's a conference Abstract written by my graduate student Chandra Moffat. It concerns the suitability of a European insect (gall wasp) to control a pest plant (hawkweed) that's been introduced to North Amer-

---

writing need not be quite so lean, there is room for debate about footnotes and other parentheticals. Sword (2012) deplores the frequent use of footnotes as "the spilled sewage of excessive marginalia," but I chose to use them to share some reasons that thinking about writing needn't be dull. Some readers have loved my footnotes, and some have hated them. I apologize if you've hated them.

ica. The central question is whether any collection of the wasp would do, or whether wasps collected from different places or different plant hosts vary in ways that would affect their usefulness for biological control. A revision deliberately targeting length took it from the "draft" version at 394 words to the "short" version at 276 words: thirty percent shorter, and noticeably easier to read (even if you don't understand the technical content). Here they are:

Is cryptic diversity in the *Pilosella* leaf-gall wasp associated with geography, host plant, or *Wolbachia* infection?

Draft version:

Biological control programs require accurate assessments of the host-range of agents prior to introduction. Discoveries of cryptic genetic variation of specialist herbivorous insects are being made at a dramatic rate, and have appeared in a few weed biological control systems, but only post-introduction. Here, we present the first study, to our knowledge, showing cryptic genetic differentiation in a candidate weed biocontrol agent prior to release.

The gall wasp *Aulacidea pilosellae* is a candidate biocontrol agent for multiple species of European hawkweeds (*Pilosella*) that are invasive in North America. Preliminary surveys in the wasp's native range in Europe suggested that it has both Northern and Southern biotypes, which appear to differ in host range, voltinism, and reproductive mode. We performed thorough and widespread surveys on multiple host species in four distinct geographic areas and sequenced three gene regions (CO1, 28S, IT-SII) to determine (i) whether there was any genetic evidence supporting the hypothesis of multiple biotypes, (ii) whether variation was based on geographic separation (as we predicted) or on host-plant association, and (iii) whether any individuals were infected with the bacterial endosymbiont *Wolbachia*, which is known to alter a number of life history characteristics in Hymenoptera.

We found considerable genetic divergence among populations of *A. pilosellae*, providing genetic support for the hypothesis that this species has multiple biotypes. This variation was found in the CO1 region, but not in the nuclear regions sequenced. A MrBAYES 50% majority-rule consensus phylogeny organized the populations sampled into three distinct lineages. While we predicted that any variation would be based on geographic separation of populations, the lineages clustered our populations into host-associated groups. Individuals collected from the host *P. officinarum* in the Northern Range clustered with individuals collected from the same host in the Southern Range, rather than with individuals from nearby collection sites which were found on different host species. Finally, only a few populations tested positive for infection with the endosymbiont *Wolbachia*: all of these were sampled from populations clustered in a single lineage. Our results have major implications for the biocontrol of *Pilosella* hawkweeds. Prior to our study, differences in biology between biotypes were attributed to geography. In contrast, we found that biotypes correspond to host association. These results demonstrate the value of genetic typing among conspecific populations of biocontrol agents to better define host-associations and reduce the risk of non-target attack prior to agent release.

Short version:

Biological control programs require accurate assessments of the host-range of agents before introduction. It is increasingly clear that cryptic, host-associated genetic variation is common in specialist herbivorous insects, but for weed biocontrol agents genetic structure has been reported only for a few post-introduction systems. We demonstrate, for the first time, cryptic genetic differentiation in a candidate agent prior to its release.

The gall wasp *Aulacidea pilosellae* is a candidate agent for biocontrol of several European hawkweeds (*Pilosella*) that are invasive in North America. Preliminary surveys in the wasp's native range in Europe suggested that it has Northern and Southern biotypes differing in voltinism, host range, and reproductive mode. We surveyed wasps on multiple host species in four geographic areas and sequenced three gene regions (CO1, 28S, ITSII) to determine (i) whether there was genetic evidence for multiple biotypes, and (ii) whether genetic variation was associated with geographic separation, host-plant association, and/or infection with the bacterial endosymbiont *Wolbachia* (known to alter reproductive mode in Hymenoptera).

We found considerable genetic divergence among *A. pilosellae* populations, supporting the hypothesis of multiple biotypes, although this reflected variation only in CO1 and not the nuclear regions. A 50%

majority-rule consensus phylogeny suggested three distinct lineages, which primarily corresponded with host association, not geography. *Wolbachia* infection was found in several populations, all of which were grouped in one host-associated lineage. Our results have important implications for hawkweed biocontrol, because they suggest that different wasp accessions will better target different members of the invasive *Pilosella* complex. Our results demonstrate the value of genetic typing in source populations of biocontrol agents, in order to define host associations and reduce the risk of non-target attack.

Note that you might not have produced the same "short" version as Chandra did: there are always choices, and your version might have stressed a slightly different story or told that story a bit differently.

It's important to point out that I didn't choose this example because Chandra is a terrible writer producing unusually bloated drafts. In fact, this is a typical example of a good writer working at her craft. As most writers should, she chose momentum over polish in the first draft, then improved what she'd written during self-revision (see chapter 21) and by taking advantage of friendly review (see chapter 22).

A final lesson here is that each writer has their own habits. Chandra's draft Abstract, for example, was relatively rich in redundancy and roundabout phrases, but largely avoided excess hedging. Your own writing may be different: perhaps you have a fondness for parentheticals, or metadiscourse, or padding phrases. Don't be ashamed of these habits— we all have them. Instead, think of familiarity with your habits as an easy way to find opportunities to streamline your text.

Chapter Summary

- Brevity in writing benefits readers, publishers, and writers.
- Most writers have habits that work against brevity—for instance, use of the passive voice, roundabout phrases, empty modifiers, and excessive hedging and metadiscourse.
- Because momentum is so important in writing the first draft, achieving brevity is a matter for revision. Dramatic reductions in length should be routine.

Exercises

1. Choose a passage (of about five hundred words) from a first draft you've recently written. Rewrite it, aiming for a twenty-five-percent reduction in character count without the loss of important information. Tally your changes against the list of eleven common text-bulgers (plus a twelfth category, "other"). Which habits present you with the best opportunities to improve brevity? Could you reduce your passage further?

2. Choose a similar-length passage from a paper recently published in your field, and attempt to edit it for further brevity. How much were you able to cut? Were there any places where additional length would have helped the reader?

3. Ask a classmate or colleague to do Exercise 2 with the same passage you chose. Compare your shortened versions, and discuss different editing choices you made.

PART V

||||||||||||||||||||||||

Revision

An unfortunate aspect of undergraduate science education is that almost all the writing involved is done to a relatively short deadline (a semester at most), and is undertaken individually, handed in, and graded. Students usually have little opportunity for repeated revision and polishing of their work. Even the few scientific writing courses available rarely feature more than one or two rounds of revision after comments from a single instructor.

This bears little resemblance to the way scientists write in the real world. The production of a complete manuscript draft is a notable landmark, but one closer to the beginning of the composition process than the end. Every paper I write gets overhauled repeatedly, going through multiple rounds of revision before anyone else ever sees it, followed by many more rounds in response to comments from friends and colleagues and then from peer reviewers and editors. Dozens of drafts are routine, spread over many months (occasionally years).

You can't learn to write better without learning to revise better. Superficially, revising might seem a matter of finding the bad parts of your drafts and fixing them, but for good psychological reasons, that isn't as straightforward as it sounds. You can greatly improve your writing craft with careful attention to how you revise, to your behavior while you're revising, and to the way you think about and interact with the people who help you by commenting on your manuscripts. Revision is a long and painstaking process, but it's an inextricable part of writing—and an opportunity to have your writing profit from your own hard second look and the responses of others.

Revision, of course, is important to all writing, not just scientific writing. Most of the advice in chapter 21 is therefore universal. In the remainder of Part V, I focus on the journal paper and the process of journal publication. It won't be difficult, though, for you to extrapolate much of that advice to help you with other writing forms.

TWENTY-ONE

||||||||||||||||||||||||||||||||||||||||||||||||||||||||||||||||||||||||||||||||||||||||||||||||||||||||||||||||||||||||||||||||||||||||

Self-Revision

You've just written the final sentence of your Conclusions (or whatever you tend to do last). There it is, on the screen in front of you: a complete manuscript. You have to admire your deft writing touches (if you do say so yourself). It even looks really good, thanks to the wonders of word processing—that's a lovely font you've chosen, and the italics make the subheads stand out effectively. Is it time to share your accomplishment with the world at large?

No! A few writers produce "first drafts" that are ready for public consumption (chapter 2), but most of us will never join their ranks. Nearly everything written by nearly every scientist goes through three more stages before it's ready for its readers: self-revision, friendly review (comments from colleagues and friends), and finally formal review (journal peer review or its equivalent). These stages have distinct functions and follow different processes, so I treat them in separate chapters.

Friendly and formal reviews (chapters 22 and 23) are invaluable tools for improving the clarity and quality of your writing. However, the kindness and patience of your reviewers are not inexhaustible resources. If you send them manuscripts that aren't the best-polished you can manage, you'll find they become less willing to help you with your next offering. If you've ever read a manuscript for someone else, and found yourself muttering under your breath "How is it *my* job to catch such obvious mistakes?," then you know exactly what I mean. And if you haven't yet muttered along those lines, before long you will.

So before you ask anyone else to help you edit your draft manuscript, you owe it to them and to yourself to do as much self-revision as you can. If you do this well, you'll send out something that reviewers will be happy to read, happy to help you improve further—and, ultimately, happy to accept for publication.

When Not to Self-Revise

While it's important to know when (and how) to self-revise, it's also important to know when *not* to. In particular, you should avoid self-revision **while you are writing your first draft** and also **immediately after you've completed that draft.** The first point was a major message of chapter 6; remember to storm the beach. The second point is equally important. Avoid the temptation to start self-revision as soon as you've finished a draft. Actually, you might not face this temptation—upon completing the first draft, you may be so sick of the project that you can't bear to look at it for a while. But if you *are* fired up and ready to dive into self-revision right away, fight the urge. Instead, put the draft away and don't think about it for a week or so. Your draft won't change in that week, but your ability to look at it critically will. In particular, the major challenge of self-revision is to see the text as a reader rather than as its writer—a psychological trick that is difficult at the best of times, but nearly impossible if the draft is too fresh in your mind. Meanwhile, if you're champing at the bit to keep writing, that's great—just turn to another piece of writing. Momentum, after all, is too precious to waste.

Taking Self-Revision Seriously

Beginning writers—and even senior ones—often struggle with self-revision in an interesting way: they *fail* to struggle with it. Instead, they give their draft a quick once-over, fix the inevitable grammatical errors and typos, and pronounce it much improved. Writers for whom this will suffice are rare (and unlikely to be reading this book). For the rest of us, self-revision doesn't mean a little polishing. Instead, it means grappling seriously with every sentence and every word in your draft. It means critical self-destruction and reconstruction. It means fixing material that doesn't work or removing material that doesn't fit, no matter how much blood, sweat, and tears it took to produce it.

Deleting material you toiled over can be heartbreaking, but delete you must. "Murder your darlings," Sir Arthur Quiller-Couch (1916) famously advised. Quiller-Couch[1] was referring in particular to excesses of style. More completely, his advice ran, "Whenever you feel an impulse to perpetrate a piece of exceptionally fine writing, obey it—wholeheartedly—and delete it before sending your manuscript to press. *Murder your darlings*." (Whether there might be a place in scientific writing for *some* "exceptionally fine style" is something I take up in chapter 28.)

Quiller-Couch's advice applies to content just as aptly as to style. You'll find that it's routine to labor over a paragraph, a figure, or an analysis, only to find upon critical rereading that it just doesn't belong. Perhaps the story you're telling has shifted during writing, or what seemed relevant in your mind seems unnecessary on paper. As an example, in a recent paper (Heard and Kitts 2012) I dealt with the impact of an herbivorous insect on two species of goldenrods. In the Introduction I brought up the notions of resistance to, and tolerance of, herbivory. In a nutshell, *resistance* means plants fighting off insect attack, and *tolerance* means plants growing despite attack. But each concept is more slippery than that, and I found myself providing detailed definitions and adding a figure and half a dozen citations. At this point I converted the passage to an Appendix, which grew to several pages of text with even more citations. All this took about three days of solid writing effort—but during self-revision I realized that the material was tangential to the manuscript, and that some fine reviews of resistance and tolerance were already in the literature. With an anguished sob, I deleted the whole thing, and not a word survives in the published manuscript. When you find a passage like this in your own work, take it out!

If you waver in your ruthlessness, hesitating as your finger approaches the Delete key, consider a bit of self-deception. This is how I soften the blow: rather than just deleting excess material, I cut-and-paste it to a

---

[1] The deliciously-named Sir Arthur Quiller-Couch was a Cornish novelist, poet, critic, and anthologist. His 1916 *On the Art of Writing* doesn't stand up terribly well today, but his incitement to "murder your darlings" is widely quoted. *Murder Your Darlings* is also the name of an Ohio metal-punk band who describe their music as "displaying the darkness of the downtrodden working man and an entire generation of Southern rockers, kick[ing] it up with Northern hardcore-punk nihilism and finish[ing] with a touch of Midwest noise-rock buzzsaw gravy" (http://www.murderyourdarlings.com/index.html). I suspect this would have puzzled Sir Arthur as much as it delights me.

separate document that I name "cuts_to_maybe_restore." I can't re-member ever actually restoring something from one of my "cuts" files, but somehow the theoretical possibility of resurrection makes it easier for me to murder my own darlings.

Getting Out of Your Head and Into Your Reader's

The key to effective self-revision is a mental trick that's probably the single most important piece of the writing craft: getting out of your own head, and into the reader's. In order to assess your writing—to find where your bid for telepathic clarity fails, and decide how to fix it—you need to read your draft **as if you were the reader you're trying to reach**. Of course your draft is perfectly clear to *you*: you wrote it, and you know what you meant to say. But communicating telepathically with your *own* mind isn't an impressive trick! To really evaluate your draft you need to achieve, or at least simulate, a mental state in which you have access only to the information a reader does: to your text, and to whatever back-ground knowledge you can legitimately assume from your intended au-dience. This means forgetting what you meant to say, and forgetting all those things you know but didn't put in the text. For instance, "reader-you" should get confused by a pronoun that lacks an obvious anteced-ent, even though "writer-you" knows perfectly well what's intended.

We can call the necessary mental trick "reader simulation," and it's a specific case of a more general mental ability that psychologists call hav-ing a "theory of mind." Theory of mind is your ability to simulate (or work out) what another person is thinking, independently of your own mental state. This includes the ability to realize that the other person lacks knowledge that you have (Box 21.1).

Box 21.1 Theory of mind

Theory of mind is a bit abstract and can be hard to grasp without a con-crete example. There's a classic experiment that makes it concrete. Imag-ine that I tell you this story:

Alice and Bai were watching TV together. Bai watched Alice put the TV remote on the end table, and then went to the kitchen to get more snacks. While Bai was gone, Alice moved the TV remote to the bookcase. After returning to the room, Bai wanted to change the channel.

I diagnose your theory of mind by asking you where Bai looked for the remote. If you say "on the bookcase," you fail to realize that Bai lacks knowledge that you possess (that Alice moved the remote while he was gone). But if you say "on the end table" you have constructed a simulation of Bai's mental state, and this simulation differs from your own. You have an effective theory of mind.

In revision, you need this ability to simulate the reader's mental state independently of your own, in order to see your text as a reader will see it. You need to set aside your own knowledge of what you meant to say, realizing that it's inaccessible to your reader—knowledge equivalent to your realization, unavailable to Bai, that Alice moved the remote.

Achieving reader simulation is not a trivial task. Deploying one's theory of mind takes effort, and we tend not to bother without a conscious decision that doing so is necessary. You know this if you have a friend who loves oysters or Brussels sprouts or peaty Scotch, and keeps offering them to you no matter how many times you explain you don't like them. It doesn't mean your friend has a defective theory of mind; rather, without conscious resolve otherwise, most of us tend to project our own thinking onto others. The same conscious resolve is necessary during self-revision.

Fortunately, with practice you can improve your ability to slip deliberately into reader simulation. In the meantime, there are a number of simple techniques you can use to help keep you thinking as the reader, rather than as you:

- **Read for self-revision in a different place or time than you wrote your draft**. Memory is strongly keyed to the context of learning, so it's much easier to remember something if you are exposed to the same environment in which you learned it (Godden and Baddeley 1975). For instance, facts painstakingly memorized while studying in a coffee bar may be distressingly elusive when you're writing a final exam in a gymnasium—unless you're lucky enough to catch a whiff of your instructor's cappuccino passing by. In the educational and psychological literature, this phenomenon is normally seen as something that can be exploited to improve memory, but in writing you can exploit it in reverse as a way to help *disrupt* memory. If you wrote your draft on weekday afternoons in your office, you may want to begin self-revision on weekend mornings in the library or in your child's treehouse. The difference in sights, smells, and sounds will make it easier for you to leave your own head and get into the reader's. And if every now and again you ask yourself what on earth you're doing working in the "wrong" place or at the "wrong" time, answering the question can help with your conscious focus on reader simulation.

- **Read for self-revision at the time of day you think least clearly**. All of us have circadian rhythms in physiology, including in aspects of mental performance (Carrier and Monk 2000). You probably know whether you're an early bird like me (pretty smart in the morning, but dumber than a rock by suppertime) or a night owl whose thinking is muddled before noon. You probably even take advantage of this by taking on mentally demanding tasks at your clearest-thinking time of day. But achieving reader simulation may be easier if you turn this logic on its head, and read for self-revision when you *don't* think so clearly. This isn't about trying to simulate stupid readers; rather, you are looking to counterbalance your overfamiliarity with what you meant to say with a bit of useful mental fog. If your draft is clear to you even when you're not thinking your best, great—and if it's not, you've found something to fix.

- **Convert your draft to an unfamiliar font or medium**. A simple trick to achieve the unfamiliarity you're after is to change to a font different from the one you usually write in. If the strangeness of the font gives you an odd sensation of reading someone else's words, you're succeeding at reader simulation. Similarly, if you normally read on the computer screen, print out your draft and read on paper.

- **Read your draft out loud**. When you read familiar text quietly to yourself, it's very easy to read what you meant to write, not what you actually wrote. It's much harder to make the same mistake when you hear your work read out loud. Mind you, if you have oratorical or dramatic experience, this is a good time to forget it—your aim is not to make the text sound its best, but rather to lay bare its flaws. Read what you wrote as simply as you can, with each word pronounced and each punctuation mark interpreted as it lies. Listen for awkwardness, repetition, and unclear meaning. As you find these issues, mark them for later and keep reading aloud.

- **Post a reminder**. The four techniques above are aimed at weakening your unconscious tendency to think as the writer. But you should not neglect the converse: opportunities to strengthen your conscious decision to think as the reader. Any time that you can explicitly ask yourself "How would a reader take this passage?," you have a chance to assess your progress toward crystal clarity. The problem is that it's difficult to maintain conscious reader-simulation, and natural instead to slip back into being writer-you. So post a reminder: make a sign that says "Be the Reader" in large, friendly letters, and hang it directly over your workspace. Move it around, too, so it won't get too familiar and recede into your mental background. Print it in the margins of your draft, if that helps. Each time you notice it, reinforce your conscious reader-simulation.

- **Target typical problems of writer familiarity.** Some particular writing problems are likely to arise, or to be overlooked, precisely because you know what you meant to say. In self-revision, you can target these deliberately. Most of these have been covered in more detail in earlier chapters, so brief reminders will suffice here. Problems closely linked to writer familiarity include:

  - **Unclear pronoun antecedents**. Check every pronoun (especially the demonstrative *this*, *that*, *these*, and *those*), asking "if I knew nothing more than the words on paper say, what could I reasonably think this pronoun refers to?" If there is more than one answer, or if the answer differs from what you intended when you wrote, fix the problem.

  - **Misaligned topic sentences**. You know how your argument flows, and how each paragraph is intended to contribute. But your reader may not. For each paragraph, ask whether the topic sentence clearly signals the content to follow.

  - **Missing transitions**. It may be obvious to you why topic B immediately follows topic A, but you've been thinking about the relationship between A and B for years. Make sure you've made transitions between ideas, paragraphs, and sections smooth for the reader.

- **Assumed knowledge**. It's surprisingly easy to leave important information out of your draft. I once read a manuscript in which the author supplied the Latin name of the study organism—but neglected to tell me that it was a plant, something that mattered quite a bit but didn't become clear from the context for some time. Particular trouble spots for the assumed-knowledge problem are the Abstract, the beginning and end of the Introduction, and the Methods, so scan these especially carefully and ask yourself if your intended audience can be counted on to know everything your text assumes they do.

Decomposing the Process

The sort of major overhaul that I've called for may sound like a daunting task. It can be even more daunting—and unlikely to succeed—if you just start doing it. It's a much better idea to break the process down into a series of smaller steps that you can tackle one at a time. This has at least two big advantages. First, bite-size pieces of the self-revision project are psychologically easier to begin (chapter 5), and completing the first one or two can give you momentum that makes it easier to keep going (chapter 6). Second, most of us greatly overestimate our multitasking ability (e.g., Bowman et al. 2010, Wang et al. 2012). If you are reading to look for overall logical flow, for example, your mind tends to skip over spelling errors, citation accuracy, or wordiness. Successfully targeting one of the latter, on the other hand, pretty much requires a willful disregard of the former.

Multiple rounds of self-revision, each with a specific aim in terms of changes to content or style, are thus the way to go. But how many rounds, with how many aims? You will have to discover what works best for you, but earlier in your writing career the number is likely to be larger. In fact, you may be startled to discover just how large. With experience you may find you can combine some, but you'll probably always go through enough rounds of self-revision to become heartily sick of what you've written. Don't worry. We all get sick of what we've written. As a starting point, I recommend you try at least five rounds of self-revision, with distinct aims as follows:

- **Revision for content**. The first step is to ask yourself again what story you're trying to tell (chapter 7). Once you know the answer to that (and it may have changed since you began work on the draft), then you can ask whether all the pieces of your draft really belong. If a paragraph, a figure, or even a sentence isn't necessary for the reader to understand the story you're telling, take it out. No less important, although usually easier in terms of self-motivation, is to ask what elements of your story might be missing and to put them in. Finally, make sure the story promised in your Introduction is the same one delivered in your Discussion. Read the first and last few sentences of your Introduction, and skip directly to the last paragraph of the Discussion (or Conclusions). Is the major point presented the same in each place? If not, adjust.

  This should be your first round of revision, because you will nearly always remove material at this step—and you might as well remove it before you waste effort polishing it. The order of the remaining rounds is up to you.

- **Revision(s) for problems of writer familiarity.** Now that the basic content is fixed, use all the tricks outlined above to put yourself into the reader's mind and read to assess the crystal clarity you're aiming for. You may be able to do this in one pass, or you might want to break this down into several substeps: for instance, a round strictly to check pronoun antecedents and one strictly to check topic sentences.

- **Revision for brevity.** Even if you wrote your first draft with brevity firmly in mind (chapter 20), it's virtually certain that you can do better. So you should always work to shorten your first draft. Most writers, even very experienced ones, make quite substantial cuts in self-revision, although of course the time to stop is when further cuts would make your writing cryptic. In my experience most writers should target at least a twenty percent reduction in overall length from the first draft. In time, you'll learn whether your own writing style needs more ruthless tightening, or whether you can get away with a bit less.

  If you find revising for brevity difficult or tedious, think of it as a game you're playing with yourself. Call it "character-count limbo," and ask "how low can I go?" Or set yourself some targets with promised rewards: a doughnut if you can hit twenty thousand characters or a walk in the woods if you can cut eight hundred words. Remember as you do this that excising material isn't evidence of unsuccessful writing—it's evidence of successful revision. So celebrate your cuts, and don't feel bad about the need to make them.

- **Revision(s) for citations.** This is deadly dull, but has to be done: check your use of literature citations. Do you have all the citations you need? Do you need all the citations you have? Do the cited papers actually say what you think they do? Have you used a consistent format? Does every citation in your text appear in your References section, and vice versa? (In my own self-revision, I have a separate round just for that last issue.) Use of bibliographic software while you write can reduce the number of citation problems, but it can't entirely replace your sharp eye.

- **Revision(s) for your personal bad habits.** Finally (or at least in my own self-revision this is the final round), you should do a special focused search for bad writing habits you just can't seem to stamp out. All of us have them. I use parentheses, for instance, as if I'd gotten an irresistible deal on a bulk purchase of water-damaged ones. My only way of fighting this seems to be a special round of self-revision in which I do nothing but search-and-destroy parentheses. Perhaps you are in love with the semicolon, can't suppress use of the passive voice, or overuse a word like "utilize" or "manifest." As you gain practice writing, you will develop a list of your involuntary writing tics, and you'll know what to look for in your own drafts.

A Final Polish

Once you've made it through all the steps outlined above, take a deep breath. Better yet, take a breather—put the manuscript away again, at least for a day or two. Then give it a final read-through with a fresh eye. It's almost certain that you'll find a few problems that you introduced while fixing other ones. But don't let fear of the next step—letting a friendly reviewer see the manuscript you've produced—trap you in an endless cycle of ever-tinier revisions. Your job in self-revision is not to achieve absolute perfection by yourself, but rather to make it as easy as possible for friendly and formal reviewers to help you improve further. It's now time to put your work out there.

Chapter Summary

- Every manuscript requires serious, extensive self-revision.
- The major challenge of self-revision is "reader simulation": forgetting what you know, and seeing the draft as a reader would.
- Reader simulation is eased by reducing familiarity with what you've written. Put the manuscript aside for a while, then work with it in a new place, a new medium, a new font, etc.
- Because we don't multitask well, self-revision works best with multiple rounds targeted at different problems.

TWENTY-TWO

‖‖‖

Friendly Review

Even the best scientific writers rarely produce excellent work in a vacuum. No matter how good you become at reader simulation, there's no substitute for the keen eye of an actual reader—one who genuinely doesn't know what you meant to say, and can react to what you've written the way your intended audience ultimately will. While most (if not all) of your papers will undergo formal review after you submit them to a journal (chapter 23), you should expose everything you write to actual readers before you get to the submission stage. This is what I call "friendly review": the stage at which you solicit comments from a friend or colleague who will read your work with an eye to making it better.

Whom Should You Ask to Review?

In selecting a friendly reviewer, you will of course want someone willing to review your work, but also someone able to provide cogent criticism and suggestions that lead you to real improvements. There can be some tension between these two criteria. For example, those closest to you are likely to be most willing to read your work—your labmates, friends from your grad-school cohort, even a spouse or family member. However, these people may not be best placed to provide helpful reviews, because they know you and your work too well. Your ideal friendly reviewer will represent your target audience, and unless your manuscript has an extremely narrow focus, you presumably will want it to be read and appreciated by readers who work with other study systems or in subfields other than your own. Therefore, you might ask for friendly review from someone you've collaborated with on *other* topics, someone in another research group or another department, or someone you've met at a conference. This is not to say you should neglect writing help from close at hand, though: a good compromise is to seek friendly review first from a close colleague, and then, after a round of revisions, to ask for a second friendly review by someone a bit further afield.

Once you've been through the friendly review stage a few times, you're likely to strike up reviewing relationships with colleagues who provide particularly helpful comments on your work. While it can be tempting to use reviewers who correct your spelling and grammar while saying only good things about the bones of your manuscript, you benefit most from the ones who are genuinely critical and unafraid to raise issues requiring hard work of you as the author. Cultivate these people, thank them graciously and repeatedly, and be quick to help them with their own work in return.

Making the Reviewing Job Easy

When you ask someone to review your writing, you're asking them for a substantial favor, given the incessant demands on everyone to review for journals, granting agencies, and so on. When you approach a potential friendly reviewer, you can entice them to agree and to stick with the job in three ways. First, you can suggest—presuming that it's true—that you think they'll be interested in your manuscript. Second, you can offer to return the favor with a friendly review of their work. Third, you can make the favor as small as possible—by making reviewing easy. After all, you'll want to use a good friendly reviewer again, but you'll encounter reluctance if you made the job needlessly difficult the last time around. At a minimum, you should:

- **Make every reasonable effort to polish your draft before sending it**. Good friendly reviewers are too valuable to discourage by sending them work full of problems you should have spotted yourself. Thus, friendly review should follow thorough self-revision (chapter 21), including careful proofreading for grammar, spelling, and the like. There are two exceptions to this guideline. First, it can sometimes be helpful to seek advice very early in a writing project on a fundamental issue such as the manuscript's overall structure or focus. In such a case, you can approach a friendly reviewer with a specific question:

for instance, "Could you skim this and see whether you think the order of the subsections makes sense? Just ignore everything else, please." Be explicit that you are delivering a manuscript that's fragmentary or unpolished, and that you don't need a thorough read. This shouldn't be a routine step for every manuscript—save it for cases where you have a real puzzler on your hands. Second, some writers negotiate an agreement for mutual friendly review with a peer, or a writing group of peers, in which the favor each writer is asking (early review of a rough draft) will be paid back in kind. Such an arrangement may be particularly useful to you if you struggle with structural issues such as sequencing material or packaging results into a clear story.

- **Allow a reasonable time for the review**. Assume that your reviewer is at least as busy as you, and recognize that your manuscript is quite naturally a lower priority for their career than it is for yours. It's usually reasonable to ask for comments within three to four weeks. If you have a deadline closer than that, it's your problem, not your reviewer's—next time, anticipate deadlines and build time into your writing process. Requests for faster review should be extraordinary, and probably accompanied by an extraordinary inducement. (Chocolate often works well.)

- **Offer the choice of an electronic or paper manuscript**. Some scientists like to read manuscripts on paper and scribble comments in pencil. Humor them! If you offer the manuscript electronically, provide a standard file format, but also one that's easily marked up. (Adobe PDF format, while universally readable, is awkward to edit.) A good choice is to provide a copy in your field's most common word-processing package (Word, LaTeX, or the like) along with another copy in PDF.

- **Double space, and use page and line numbers**. Line numbers are especially helpful, as they let a reviewer refer easily to specific features of the manuscript: "What you say at line 321 seems to contradict your earlier statement at 162." No one enjoys counting by hand to refer to the "fourteenth line of the second paragraph on the seventh page."

- **Ask specific, concrete questions**. You can increase the review's utility without significantly increasing the reviewing burden if you draw the reviewer's attention to a limited number of areas you'd especially like help with. For example, you might mention that you are not sure if a piece of terminology is clear enough, that you are worried about the right level of detail in the Methods, or that you feel the Discussion is overly long and would like advice on what to cut.

- **Provide appropriate, polite reminders**. A single, understated reminder by e-mail after two or three weeks have gone by is perfectly appropriate, as is following up again two weeks after your first reminder. But don't nag—reminders too early or too often come across as demanding. It's best to decorate your reminder with an offer of help—for instance, "I hope you were able to open the file; please let me know if you'd prefer a different format." Finally, if your manuscript gets pushed down your reviewer's priority list by something truly important, be gracious about it.

How to Read a Review

Early in my writing career, I knew I had a lot to learn. I knew I wasn't very good at maintaining momentum to get through first drafts; I knew I had a weakness for complex sentences and digressions; and in my more honest moments, I could have provided a long list of more problem areas. But I was completely unaware that I didn't know how to read a review. In fact, I had no idea that this was something a writer could be good or bad at. So how can you be a good reader of your reviews?

When you receive a review, read it through right away. But then stop, **and do nothing else.** Don't read it again, don't scribble notes on it, don't begin to make revisions, don't vent to a friend about how your reviewer misunderstood you—do nothing. Instead, set the review aside for at least a day or two. Few, if any, of us are capable of reacting calmly and constructively to criticism of our work (however spot-on it may be) on first reading.

Once your reaction has had a chance to season, pull out the review again. Now go through it and divide up the comments into three categories. You'll deal differently with each category in revision. Mark as **category one** comments that have obviously identified clear problems with straightforward solutions—say, grammar mistakes, reordering of sections, or a new title. Mark as **category two** comments that seem to be on to something, but for which you don't (yet) see an obvious path forward. Perhaps the reviewer read your work from an angle you hadn't thought of, suggested a new analysis that you're not sure you can do, or suggested an alternative statistical test that you aren't sure is appropriate for your data. Finally, mark as **category three** comments to which your reaction is "that idiot didn't get my point at all/didn't bother to read/etc." There will almost always be comments like this, and they will make you angry, but they're important and when considered properly can help you make major improvements to your writing.

Category-one comments are (by definition) easy to deal with, and when you begin to revise, you'll address these right away. Doing so will help ease you back into work on a manuscript that you've likely put aside for a while, and it will reacquaint you with the text so you'll have an easier time with categories two and three.

Category-two comments involve harder thought and more work. They'll typically involve points of science rather than just wording or arrangement, and deciding whether substantive changes are merited (and what those changes should be) may require learning new techniques, reading more literature, or thinking about your work in a different way. Two quite natural reactions can lead you to resist dealing adequately with a category-two comment. First, you may resist change simply because your original analysis or approach seemed right to you when you did it, and still does. You may be correct about this, but you may also be reluctant to move out of a comfort zone or to recognize imperfection in your earlier thinking. You should acknowledge these reactions and therefore set a high bar for convincing yourself to stick with your old approach over your reviewer's suggestion. Second, you may resist a substantive change simply because carrying it out will take a lot of time and labor, and you thought you were finished with the manuscript. Because word-processing software and laser printers can make any draft look just as tidy as a published paper, it can be psychologically difficult to accept that your work actually needs major change. Odds are good, though, that if your friendly reviewer has suggested a substantive change, formal reviewers will do the same—and you might as well tackle the issue now while it doesn't imperil acceptance of a submitted manuscript. When revising to deal with category-two comments, it can often be useful to engage in some discussion with your reviewer, to ask for clarification or for further suggestions. Don't, however, *argue*—at least, not if you want future reviews from the same person.

Category-three comments are the ones that early-career writers mishandle most often. Here you've decided that your reviewer got it wrong, whether from inexperience in your field, superficial or careless reading, or innate stupidity. In fact, most of the time, category-three comments identify areas in which your attempt at crystal-clear writing has failed. If you reread your text with an open mind, you'll usually discover that your reviewer was led astray because you didn't make your point clearly, or you buried it in uninteresting or confusing text. And even if you were right—what of it? It's your job to make your writing clear even for a reader who is in a rush, distracted, careless, or even (gasp) not quite as brilliant as you. So while a category-three comment may annoy you at first, rather than declaring angrily "that's not what I meant at all!," recognize an opportunity to understand what the reader is experiencing in your text. If your reader comes to a conclusion that differs from what you intended, that's something you should want to fix. Taking this perspective sets responsibility for reader misunderstanding where it should almost always be: on you as the writer.

A Special Case: Serial Friendly Review

In some cases, the friendly review process may be more protracted as you send multiple drafts of the same manuscript to the same reviewer. Such serial friendly review occurs most commonly when graduate students send repeated versions of thesis chapters or manuscripts to their supervisors. When you seek serial friendly review, there are a few additional practices that can help maintain the good nature of your reviewer.

Unlike the usual friendly-review case, a serial friendly reviewer will see your revisions—and see them soon enough after commenting that they will likely remember much of what they said. Furthermore, they will often look specifically to see how you dealt with their comments in the new version. You can make this easy for them by offering to provide a brief document outlining the changes you made and how they address the reviewer's concerns. (This "response to reviews" is even more critical in the context of revising manuscripts following formal review, and is dealt with in detail in chapter 24.) Respond to every substantive comment: if you made a recommended change, say so; and if you did not, then explain what other change you made that solves the problem your reviewer identified. What you should *never* do is ignore a comment. Nothing raises a supervisor's blood pressure more quickly than correcting the same apparent mistake in a third consecutive draft!

Finally, a serial friendly reviewer may well see your *next* manuscript after handling repeated drafts of this one. In this case, you can build mountains of goodwill by refraining in a second manuscript from mistakes corrected during revision of the first. While this sounds obvious, it's not trivial to accomplish, because the very fact that you made the mistake suggests it may be one to which you're prone. Therefore, before sending your friendly reviewer a draft of your next manuscript, you should deliberately review the revisions you made to the last one, and incorporate similar modifications unprompted. Of course, you should do this all the time, but the stakes are particularly high when you risk nettling a reader who's important to you. If it helps, make a handy list of the sins you're wont to perpetrate and use it as checklist that grows with each new review you receive.

Chapter Summary

- All manuscripts should receive "friendly review" before formal submission.
- Ease the burden on your friendly reviewers: send polished drafts formatted for easy reading, and ask concrete questions.
- Reading and responding to reviews requires careful thought. Allow time for your reaction to mellow, and be skeptical of your impulse to believe your reviewers are wrong.
- When a friendly reviewer sees repeated drafts of your work, provide a brief summary of changes with each new draft.

TWENTY-THREE

Formal Review

Once your manuscript is as polished as you (and your friendly reviewers) can make it, your next step is to submit it for potential publication. I'll assume here that you're submitting to a peer-reviewed journal, as that's by far the most important kind of outlet in the natural sciences. (Other kinds of publication are discussed in chapter 26.) The distinguishing feature of peer-reviewed publication is right there in its name: your manuscript will be read and reviewed by your peers (other working scientists). I call this "formal review," because it's an official part of the well-defined procedure by which your submitted manuscript is handled and assessed.

How Formal Review Differs from Friendly Review

Formal review has a lot in common with friendly review (chapter 22), but the two differ in at least three important ways.

First, formal reviewers are drawn from a global pool of experts, and while they'll generally know your field, they needn't know you or your previous work the way friendly reviewers do. This is a good thing, for neither will most of your intended readers. Formal reviewers can do a better job than friendly ones of representing your intended audience, and thus helping you judge your efforts to achieve crystal-clear writing. Second, nearly all journals offer formal reviewers the option of submitting their review anonymously. It's not uncommon for formal reviews to be signed anyway, especially when they come from more senior scientists. Nevertheless, the option of anonymity increases the value of formal review to both journal and author. Anonymity allows a reviewer to tell you (the author) the unvarnished truth, without fear of awkwardness or worse between you later. Of course, in a perfect world every reviewer would know how to cast criticism so it's taken as entirely constructive, and every author would be able to receive even poorly phrased criticism dispassionately—but in our world of normal humans, a reviewer's reluctance to sign a critical review is entirely reasonable[1]. The reviewer's option of anonymity therefore assures you that your formal reviewers can help you in the most direct way possible. Even the best friendly reviewers will be tempted to soft-pedal the advice you need.

Finally, a friendly reviewer reading your manuscript has only one goal in mind: making the manuscript better. Formal reviewers, on the other hand, play two roles at once. They provide comments and suggestions that can improve your manuscript, just as friendly reviewers do; but they also help the editor judge the quality of your work and thus serve as gatekeepers. At first glance, it may seem that these two roles are quite distinct: in their improvement role reviewers are serving the author, while in their gatekeeper role they are serving the journal (and are adversarial to the author, who may be denied publication). This view is too simple, however. In fact, journals benefit along with authors from the improvement role, and authors benefit along with journals from the gatekeeper role. The latter point may seem implausible (particularly if you're smarting from a recent rejection), but good reviewers and editors draw clear distinctions between a manuscript that is not publishable, one that is not publishable *yet*, and one that is not publishable *in that journal*. If your work is not publishable at all, perhaps because of a critical design flaw, you should want to know so that you can turn your efforts to a new study. Fortunately, such extreme problems are rare. If your work is not publishable *yet*, you should appreciate valuable advice about what's needed to make it so. And if your work is sound but doesn't suit

---

[1] It is true that a few reviewers abuse their anonymity to make careless, rude, personal, or otherwise unprofessional comments. However, good editors give little weight to careless reviews, and will catch and redact unprofessional language. I'm not saying that every editor is up to the job. I once submitted, to a small journal, a manuscript reporting some limited but sound and (I still think) interesting data. One reviewer wrote, anonymously, that they wouldn't have given my manuscript a passing grade in their undergraduate Introduction to Ecology course. That was it—no explanation of what was wrong, and no suggestion toward improving it. This was clearly unprofessional and an abuse of anonymity, and the editor should never have sent me the review. Luckily, no harm was done: I was too stubborn to be cowed, and I sent the manuscript to another journal, where it was accepted with a few minor revisions. And I have the story to tell over beer.

the journal you sent it to, you should want to find, and tailor your writing to, the right audience. Thinking of formal review simply as an obstacle to be overcome before publication sacrifices much available writing help.

You may object that sometimes you just need to get a manuscript published, before a critical career event such as a job or grant application or a tenure decision. At such times, you may not appreciate the delays entailed by reviewers helping you find the perfect match between manuscript and journal or pushing you to make every possible improvement. All of us have these thoughts at some point in our careers, but we can't expect journals to take our own deadlines into account. Fortunately, electronic tools have made reviewing and publication enormously quicker than they used to be, reducing the time cost of the writing help that comes with review.

The Reviewing Process

What happens to your manuscript after submission can seem rather mysterious, but it needn't be. While there are some differences among journals, nearly all follow some version of the following six-step process:

- **Manuscript check.** The submitted manuscript is checked by the journal office to make sure it's suitable to put through the review process. Its subject area must fit the journal; it must not missing be sections, figures, etc.; it must be formatted as required by the journal's *Instructions to Authors*, and its writing must be clear enough that reviewers will be able to assess content. If there are deficiencies of this kind, the manuscript is returned right away. This step is essentially a filter that saves editors and reviewers from wasting effort on manuscripts with no hope of acceptance, and it mostly catches amateurs[2], cranks, and the unforgivably sloppy.

---

[2] *The American Naturalist*, one of the top journals in ecology and evolution, gets some notably amateur submissions—possibly because of its name, which is welcomingly but misleadingly nontechnical. One would-be contributor sent, from an island in the remote South Pacific, an essay explaining that llamas have peculiar feet because they walk in deep snow a lot (which would be a

- **Editorial assignment.** The editor-in-chief reads the manuscript and either rejects it without review (see "Decoding and responding to editorial decisions," below) or else assigns it to a handling editor (sometimes called a corresponding editor, subject editor, or associate editor). Some smaller journals omit this step, but for most the volume of submissions and/or breadth of topics covered makes it impossible for a single editor to handle every manuscript. The journal *Geosciences*, for example, has an editorial board of twenty-seven handling editors, and journals with fifty or more are not unusual. Normally, the handling editor will have expertise close to, but perhaps not directly in, the field of your manuscript. Unlike the peer reviewers, the handling editor normally will be identified to you.

- **Reviewer assignment.** The handling editor reads the manuscript and either recommends rejection without review or else assigns it to two or three willing peer reviewers. You can expect the reviewers to be closer than the editor to your subdiscipline, although it's unrealistic to expect them to know the topic as well as you do. Actually, it can be quite useful to have reviewers who don't know your topic well; after all, you are probably hoping that your paper, once published, will be read and understood by people exactly like that.

- **Review.** Each peer reviewer reads your manuscript and supplies the handling editor with comments, criticisms, and suggestions in a written review. Each reviewer also recommends an editorial decision.

- **Recommendation.** The handling editor rereads the manuscript, considers the reviews, and recommends an editorial decision to the editor-in-chief. When the peer reviews all say substantially the same thing, the editor will nearly always follow their advice, but it is common for the reviewers to offer different recommendations. In this case, the handling editor will make the decision, usually reading your manuscript carefully enough to act as an additional reviewer. There is no "rule" for reconciling conflicting reviews: the majority needn't carry, and a tie needn't be resolved for (or against) the author.

- **Decision.** The editor-in-chief makes the final editorial decision and communicates it to you, along with copies of the reviewers' and handling editor's comments.

---

surprise to most llamas). Another sent a description and photograph of some unusually shaggy sheep. One reads a lot about citizen science, but I don't think this is what it means.

Favored and Disfavored Reviewers

During submission, many journals allow you (and some require you) to suggest names of appropriate reviewers for your manuscript. These are called "favored" reviewers. You may also be asked to name people you'd prefer *not* to review your work; these are called "disfavored" reviewers. Editors are not bound to use your favored names or to avoid your disfavored ones, but it's nonetheless worth your time to put some thought into both. In fact, even when a journal doesn't explicitly ask for favored and disfavored reviewers, you can and should mention these in the cover letter that accompanies your submission. Editors generally appreciate favored-reviewer lists, because they can save substantial effort in securing appropriate reviews. Even when the editor can't or doesn't use your exact suggestions, they're a guide to coming up with others.

It may seem tempting to suggest favored reviewers you think will like your manuscript (or who are friends, know your work extremely well, or tend to be uncritical). Avoid this temptation. You aren't looking for someone who will recommend acceptance of your manuscript regardless of its quality. Instead, you're looking for someone who can give you valuable criticism, so that what you eventually publish is better than you could have managed on your own. Suggest reviewers who *don't* know your work intimately, who publish work you admire, and who have lots of reviewing experience. This will often describe relatively senior or well-known figures in your field. Of course, such people get a lot of requests to review and so must decline many of them—but suggest them all the same; the worst that can happen is that the editor needs to use someone else.

The opportunity to disfavor reviewers presents an interesting conundrum. On the one hand, scientists are eminently human and if someone in the field is likely to act inappropriately reviewing your manuscript, the editor might as well know about it and avoid trouble later. Reasons for disfavoring a reviewer might include a past professional or personal disagreement or a competitive relationship between labs that might tempt a reviewer to make inappropriate use of data in your manuscript. (It would be highly unprofessional for an editor to tell a potential reviewer why, or even that, you disfavored them, so you need not worry about that.) On the other hand, disfavoring reviewers removes names from the list of experts who could help you improve your paper. In addition, an author who disfavors too many reviewers can make an editor wonder what's going on, and whether the problem might lie with the author rather than their myriad (presumed) enemies. Actually, most scientists will give a fair review even to someone they don't like very much, and so most authors only occasionally indicate disfavored reviewers. As a rule of thumb, if you find yourself wanting to list more than two disfavored names for a single submission, or any disfavored names at all for more than an occasional submission, you should probably have a frank talk with yourself about how you relate to colleagues in your field.

Decoding and Responding to Editorial Decisions

What you've submitted must, of course, come back. Beginning authors often imagine that their submission will either come back accepted (good news) or rejected (bad news), but in fact it's much more complicated. The former decision is rare, the latter comes in several flavors that differ in important ways, and most decisions are actually somewhere in between. Editorial decisions can take one of seven basic forms (although terminology may vary):

- **Editorial decline**. An "editorial decline" means that your manuscript is being returned without being sent out for peer review. (Most journals use "decline" rather than the synonymous "reject," presumably because it sounds less judgmental.) This usually means that the editor saw either a mismatch between your manuscript and the journal's scope or an extremely serious problem with your argument. If the former, simply send your manuscript elsewhere. If the latter, fix the problem, then do the same.
- **Editorial decline without prejudice**. A "decline without prejudice" (DWOP) is distinguished from a straight decline by the editor's intrigue with what's otherwise not a suitable submission. A declined manuscript can't be resubmitted to the same journal, but a DWOPed one can. However, you should understand the message clearly: the editor thinks your manuscript might have a chance in the review pro-

cess, but only if so thoroughly reworked that it's best thought of as an entirely new manuscript (and thus an entirely new submission) on the same topic.

You *can* send a DWOPed manuscript back to the same journal, but depending on the reasons for the DWOP, you may or may not want to. If those reasons involved problems with data or analyses, you'll have to fix them no matter where you submit next. In this case, the journal issuing the DWOP was your first choice initially and probably still is, and it's perfectly reasonable to resubmit there. However, if the reasons had to do with lack of journal fit or issues of manuscript content, think hard about your choice. If the editor suggests that the journal would require a very different perspective on your topic, you may have chosen the wrong venue. Rather than reinvent your manuscript, you may want to find a journal that fits it better. Or perhaps the editor suggests the manuscript might be suitable with additional data (say, literature review with meta-analysis). Here you should think about your publishing plans more broadly: if you have those data, or can get them, should you add them to the manuscript in question, or hold them for another paper? Editors will often signal their awareness of such concerns with wording along the lines of "Addition of such-and-such would be required for this manuscript to be suitable for our journal. Alternatively, the authors may want to make more modest revisions and target a more specialized journal."

The frequency of editorial declines and DWOPs has become quite high—not uncommonly twenty-five to thirty percent, and reaching as high as seventy-five percent at *Nature* and *Science*. This is a practical solution to rising submission volumes and the resulting heavy demand for the services of peer reviewers. It can be hard to secure willing reviewers, and it helps considerably if editors can make some decisions without going to that shallow well. Some authors are incensed by editorial declines, but in fact they're usually in the author's best interest. If your manuscript seems to have a fundamental flaw, or doesn't match the scope of the journal you sent it to, going through peer review won't increase its chances of acceptance. Better to get it back promptly (editorial declines often take only a few days) so you can revise and submit elsewhere without delay.

- **Decline following review.** A decline after peer review has the same consequence as an editorial decline: your submission will not be considered again by that journal. Your next steps are also the same as for editorial decline, except that you have peer reviews to help. Don't ignore the reviews on the grounds that you're sending your work elsewhere anyway. To do so is to discard free writing help, and to dismiss the good-faith effort made by the reviewers. Besides, if you ignore a review and resubmit elsewhere, the laws of karma practically guarantee that the second journal will consult the same reviewer. (If you don't believe in karma, the limited supply of available reviewers favors the same result.) Most senior scientists have at some point responded to a request to review a manuscript this way: "I reviewed this manuscript last month for Journal X and recommended rejection. It appears to be unchanged, and so is my assessment of the work. I attach a copy of my previous comments." This review is great fun to write, but very little fun to receive.

  Having your manuscript declined (whether with or without review) always smarts. However, it's not the end of the story: almost all declined manuscripts find homes, after suitable revision, in other journals. Among papers that have followed this path are classics in every field: for instance, Peter Higgs's (1964) paper predicting what would later be called the Higgs Boson, or George Akerlof's (1970) paper on economic markets with asymmetries. Both of these rejected manuscripts eventually earned their authors the Nobel Prize. There is even some evidence that manuscripts declined on first submission have, on average, greater citation impact than manuscripts that are never declined (Calcagno et al. 2012). This pattern suggests that help from an extra set of reviews, and perhaps the strong message a decline sends about revision, actually benefits authors in the long term. So stick with it.

- **Decline without prejudice following review**. DWOP following review has the same consequence and offers you the same options as does an editorial DWOP—but with additional help available from the reviews.

- **Major revision**. An editor who asks for a major revision is signalling that your manuscript is quite likely to be accepted, provided that you can make changes to address some important concerns. However, this decision doesn't guarantee eventual acceptance, especially if you don't take the word "major" seriously. You should make thorough revisions, and submit the revised manuscript along with a *Response to Reviews* (chapter 24). The editor will read your revised manuscript and may or may not send it out for further peer review (by the original reviewers or new ones). The revised manuscript then receives a new decision.

- **Minor revision**. A decision of "minor revision" is a strong signal that the editor wants to accept your manuscript. The editor believes that any criticisms raised by the reviewers have straightforward fixes that won't change your manuscript's overall story. This is still not a guarantee, but you'll have to work hard to blow your chance. As with a major revision, you will return a revised manuscript and a *Response to Reviews*. Your revised submission will probably be considered by the editor without further review. Almost always, the editor will either request further minor revision or accept your revised manuscript.

- **Accept**. In theory, your manuscript could be accepted on first submission—but this almost never happens. In real life, you'll see this decision only after a round or two of revisions following an initial "major revision" or "minor revision." Your response is simple: celebrate! Of course, you still have work to do: final formatting, uploading data to an archive, filling out copyright forms, checking proofs, and so on. However, the heavy lifting is over.

Corresponding with Editors and Reviewers

The rather formal structure of your interactions with editors and reviewers can make them seem like disembodied voices of authority from the sky. It needn't be so. Editors and reviewers are, in fact, people just like you. (There is almost certainly one just down the hall from you; at some point, if not now, there will be one directly underneath your own hair.) Like you, editors and reviewers are sensible and very nice most of the time, but grumpy every now and then. Like you, they know a fair bit about science, but not everything. Like you, they want the published literature to be of high quality, and your manuscript in particular to get better. They will work hard to those ends (and appreciate it if you thank them for it). In short, editors and reviewers are your partners, with a shared goal of having you publish high-quality papers in journals where they reach their best audience.

Thinking of editors and reviewers as partners rather than adversaries raises the possibility of corresponding with them (beyond just sending in your revised manuscript, I mean). While many authors are loath to do so, there are several circumstances in which it is entirely appropriate to contact a handling editor or even a reviewer:

- **Negotiating deadlines for revisions**. A request for revisions usually comes with a deadline for submission of the revised manuscript. Typically one to six months are allowed for revision, although deadlines as short as a week (startlingly) are possible. It is perfectly acceptable to ask for a modest extension, provided that you don't ask at the very last minute and that you have a legitimate reason for needing more time.
- **Asking for help in understanding a reviewer's or editor's comment**. Occasionally a comment may be more cryptic than the reviewer or editor intended. If the point is important, it's fine to ask for clarification. Just make sure not to come off as passive-aggressive: "I don't understand how the reviewer could have gotten that idea" is always going to be taken as carping. A reviewer who has signed a review can be contacted directly (I sign most of my own reviews precisely to invite this); otherwise, you can ask for the editor's take on the mysterious comment.
- **Requesting feedback on a potential fix**. Most of the time, it will be fairly obvious how to address a problem spotted in review. Occasionally, though, you won't be sure whether a proposed fix will be satisfactory, or which of two possibilities is better. This happens most often when a reviewer suggests something (a new analysis, say) that turns out to be impossible, but you have an alternative or two in mind. If the possible fix would require a lot of new work, or if the rest of the revision turns on which alternative is chosen, then it makes sense to ask for feedback before getting in too deep. Keep in mind that it's reasonable to ask "would approach X satisfy this objection?," but that nobody's going to answer "if I take approach X, will my manuscript be accepted?"

Contact with editors and reviewers should be by e-mail, which allows leisure in the reply. Unless invited, don't place a phone call (and never, ever show up at the editor's office or stalk a reviewer at a conference). Always be sure your correspondence is gracious, phrased constructively rather than confrontationally, and conveys gratitude. Editors and reviewers handle dozens of manuscripts a year, so you should reserve the option to contact them for times when their help will really make a difference. In my experience as an editor and reviewer, I've been contacted by the manuscript's author (outside the standard submission process) somewhere around five to ten percent of the time. This seems about right.

Finally, if your manuscript is declined, should you appeal the decision? This often seems like a good idea with emotions running high right after you've received the rejection notice (sorry, "decline" notice), but it's rarely worthwhile. Sure, it's possible for reviewers and editors to get things completely wrong, but this is rare enough that it might happen to a given author maybe once or twice in a prolific career. Even if you've genuinely been wronged, as a practical matter an appeal is unlikely to succeed, risks gaining you a reputation as troublesome, and won't be any faster than simply sending the (revised!) manuscript to another suitable journal. Better to simply move on.

Chapter Summary

- Formal reviewers play two roles: they help improve your manuscript, but they also serve as gatekeepers for the literature. Both roles benefit you as a writer.
- The review process has six steps: manuscript check, editorial assignment, reviewer assignment, review, editorial recommendation, and decision.
- A submitted manuscript may get one of seven types of decision: editorial decline, editorial decline without prejudice, decline following review, decline without prejudice following review, major revision, minor revision, or accept. Your next steps as a writer depend on the decision type.
- Correspondence with editors, and with reviewers who decline anonymity, is appropriate if handled professionally.

TWENTY-FOUR

||||||||||||||||||||||||||||||||||||||||||||||||||||||||||||||||||||||||||||||||||||||||||||||||||||||||||||||||||||||||||||||||||||||||

Revision and the "Response to Reviews"

Formal review is nearly always followed by revision of your manuscript. Your revisions have two related but not identical goals: to improve your manuscript as much as possible, and to maximize the probability of its ultimate acceptance. Skill at revision in response to formal reviews is an important part of our writing craft. It has three major elements: reading reviews effectively, conducting the revisions themselves, and drafting a "Response to Reviews" document to accompany your resubmission.

Reading Reviews: A Reprise

Reading formal reviews is not fundamentally different from reading friendly ones, so this would be a good time to reread my advice on that subject (chapter 22). But remember that formal reviewers' greater distance from you makes them particularly good stand-ins for the readers you're trying to reach. Even when your first reaction is to dismiss their complaints (and perhaps especially then), try to approach every line of a review as an opportunity to peek inside a reader's mind and to make your writing as clear as possible.

Thorough Revision

Punching the "submit" button on a journal's website is a very satisfying thing to do. Because it takes the manuscript out of your hands and lets you turn to another project, it tends to feel like the end of something—

as if the serious writing is over, and all that remains is for the review system to recognize the excellence of your work and rush it into print. This attitude is so seductive that if you don't consciously resist it, you can end up loath to do more than superficial tinkering after the reviews come back. Falling into this trap sacrifices much possible improvement. It also risks rejection, because reviewers notice when the two-thousand-word review they spent hours writing leads you to add two sentences and correct a few spelling mistakes. And yes, I've seen authors do exactly that.

The challenge, then, is to summon enthusiasm for truly major revision in the face of temptation to see the submitted manuscript as a finished product. Actually, this isn't that different from the challenge of self-revision (chapter 21): recall the difficulty, and the importance, of murdering your darlings. When considering each reviewer's comment, you may want to ask yourself, "what's the *easiest* change I could make to address this?" What you should ask yourself, though, is "what change, no matter how major, would *best* address this comment?" If a reviewer found the aim of your study unclear, perhaps your Introduction just needs an extra sentence—but on the other hand, it may need fundamental rethinking. If a reviewer wasn't convinced of a pattern you point out, maybe a figure just needs bigger symbols—but it may need a complete redesign. If a reviewer suggests an alternative interpretation, perhaps you could just insert in your Discussion, "While we cannot dismiss the possibility that thing X happens, such a hypothesis must await future work"—or you may need additional data or new analyses to close the gap in your argument. Of course, sometimes those easy changes really do suffice. I'm not suggesting pitching all your hard work in the nearest recycling bin, only being open to major change if that's what needed to address reviewer comments.

It's quite likely that nothing in this argument surprises you. Of course one should be open to major changes when problems are identified. What's interesting, though, is how easily one can believe this in general, and yet still catch oneself resisting big changes to a particular manuscript. (I've certainly done it.) The temptation to think of your submitted manuscript as complete can be powerful, and it takes deliberate thought to be sure you've overcome it.

Disagreeing with Reviewers

What if you don't agree with a reviewer's criticism? Do you have to make the suggested change anyway? In a word, no. But don't get too empowered: reviewers aren't *always* right, but it is wise to think that they *probably* are. There are two reasons for this: first, formal reviewers are often more experienced than you are; and second, you're likely to have some unconscious investment in the way you've done things, which will make you resist suggestions of change. So while any particular reviewer suggestion may indeed be wrong, you should demand of yourself a very strong argument before deciding it's so.

Assume now that after very careful thought, you have decided that you don't agree with a particular suggestion. What should you do? First, distinguish two possibilities: is your objection that the suggestion *would not make your manuscript any better*, or that the suggestion *would make your manuscript worse*? If it's the first—for instance, you did analysis X, the reviewer wants analysis Y, and they're equivalent—make the change. After all, for just a little extra work you can make the reviewer (who will pass judgement on your revised manuscript) happy. If you wish, think of this as accumulating chits you can cash when you really need to resist a suggestion.

What about the second possibility—that adopting a suggestion would be a genuinely bad idea? If you're really sure, you have four options:

- **Ignore the suggestion.** I include this option for completeness, but it's a bad idea, because your ploy is sure to be detected. Few things raise my ire as a reviewer and editor more surely than discovering that my suggestions have been completely ignored.
- **Rebut and make no change.** Explaining to the editor why a reviewer is wrong is perfectly OK, as long as you have a solid case and can argue it cogently, constructively, and politely. Because you are making no change to the manuscript, you should explain why the reviewer's incorrect opinion is *uniquely* incorrect—that is, likely to be shared by no other reader you're interested in reaching. Since this is actually quite improbable, this option is one you should use very rarely.
- **Rebut but revise.** In the vast majority of cases, you should rebut the reviewer's comment (as above), but revise your manuscript anyway.

Doing so recognizes that even though the reviewer misunderstands, this does you a favor by identifying a shortcoming in the clarity of your writing—one that could lead other readers to misunderstand in the same way. Your revision won't change the substance of what you had written, but will present it in a different, clearer way.

- **Defer to the editor**. Finally, as long as you don't do it too often (more than once for a manuscript is pushing it), it can be useful to offer the final judgment to the editor. This means saying "The reviewer wanted me to do X instead of Y, but here's why I still think Y is better. I've kept Y, but I'm happy to defer to your judgment. If you strongly prefer X, let me know and I can supply an alternative version." This shows you as cooperative, while also favoring your chances of prevailing: as long as the editor thinks you've made a reasonable case, it's unlikely you'll be asked to make the change you're offering. Of course, don't bluff. If the editor asks you to make the change after all, you'll have no option—and if that happens you should accept that, almost certainly, the reviewer and editor are right, and you are wrong.

Always remember that your paper will be strengthened by changes you make in response to review—and it may be strengthened the most by your responses to comments you don't agree with. This is why my favorite Acknowledgements section of all time goes like this:

> For critical suggestions and discussion I thank [names]. Not everyone agreed with everything but even that helped (West-Eberhard 2014).

The "Response to Reviews"

Skillful revision of the manuscript goes a long way to making acceptance likely, but there's more to the process. When you submit the revised manuscript, you'll also include a document called the "Response to Reviews" (henceforth just "Response"). This will be read by the handling editor and, if your revised manuscript is subject to further review, by the reviewers. Its immediate purpose is to outline the changes you made and how they address the reviewers' criticisms. Its broader purpose is to make it easy, if not irresistible, for the handling editor to accept your paper.

A good Response allows editors and reviewers to check quickly to see that you've responded to the comments they made on your original submission. If it's thorough, positive, and well written, they'll approach your revisions in a good mood. If it isn't thorough, they'll have to work hard to locate the material that provoked each comment, to figure out what changes you've made, and to judge whether those changes are sufficient. This will test their patience, and you don't want reviewers and editors impatient and grumpy as they decide the fate of your revision. It is therefore a mistake to think of the Response as a clerical step requiring perfunctory attention after the revisions are otherwise complete.

What should a Response contain? Most have three sections: a brief expression of thanks, a short summary of the major changes, and a longer section including the full text of all the reviewers' comments interleaved with your point-by-point replies. I'll illustrate using a real example: my coauthors' and my Response to the reviews of Halverson et al. 2008a (edited for brevity and to spare you some technical detail). Our Response began:

> *Dear Dr. Editor,*
>
> *This letter accompanies our revised manuscript "Differential attack on diploid, tetraploid, and hexaploid* Solidago altissima *by five insect herbivores." We are grateful for the reviewer's helpful comments, and hope our revision addresses them all. In particular, we've added a new analysis to test for spatial autocorrelation (Reviewer 1) and a discussion of the distinction between herbivore preference and performance (Reviewer 2). While those suggestions led us to add some new material, in response to your request for reduction in length we've edited thoroughly for brevity, and the revised manuscript is about 10% shorter.*
>
> *Below we detail the changes made in our revision. We include the text of the reviews in Roman font; our responses are in italics. References to line numbers are for the revised manuscript.*

This introductory passage signals that the authors are cooperative rather than combative, appreciative of the reviewers' comments, and eager to use them to improve. It draws attention to the most substantive changes, relating them to the reviews, and reports compliance with the (almost invariable) editorial request for reduced length. It then leads into the much longer point-by-point, establishing an easily-to-follow convention for presenting authors' replies matched to reviewers' comments. This should put editor and reviewers in a frame of mind where even before reading the revisions, they expect to recommend acceptance of the revised manuscript[1].

The bulk of the Response consists of your point-by-point replies to the reviewers' and editor's comments. When you start working on revisions, your very first step should be to cut-and-paste those comments into the document that will become your Response. This is also a good time to label the comments using the Category One/Two/Three scheme outlined in chapter 22. (Remove these labels before resubmitting.) Then, whenever you change your manuscript in response to a reviewer's comment, immediately insert an appropriate mention of that change in your growing Response. This way, you aren't stuck later trying to remember what you did in response to each comment; as a bonus, the Response doubles as a checklist for revision. Note the approximate location of each change as you make it. After finishing, you'll go back and insert line-number references so that reviewers can find each change in your manuscript.

Every comment should receive a reply, although if two reviewers make similar comments, your reply to the second one can be simple: "*Corrected; see response to Reviewer 1.*" For simple matters of grammar, spelling, formatting, and so on, not much is required: you'll have made the changes, and need only say so ("*Corrected, thanks*"). You can even group all such comments together in your Response and give the single reply "*All corrected, thanks.*" Otherwise, don't reorganize the text of the reviews, as you want to make it easy for the editor to see that you haven't left anything out.

More substantive changes will require a little more in the Response. Before getting to content, I can't overemphasize the cardinal rule that your Response should always be polite and constructive—even if you disagree with the reviewer, and even if you feel offended by a comment. Even if only the handling editor sees your Response, seeming combative

---

[1] You can blow this if the rest of your revision demonstrates that these signals were insincere. I was recently the handling editor for a revision that came with just the right Response—grateful, positive, and outlining changes that nicely dealt with the major criticisms of the original submission. I roughed out an acceptance letter, and then read the revised manuscript—which, as it turned out, didn't actually incorporate the changes that the Response said it did! I was an unhappy editor, and shortly thereafter, the authors were unhappy authors.

will only hurt your cause—but your Response will probably be sent to the reviewers, so don't write anything you wouldn't want them to read.[2]

If you've taken up a reviewer's suggestion, it's worth a sentence or two explaining the change and how it improved the paper. For instance:

> Line 289: A major point of the paper is assessing spatial structure in the distribution of diploid and hexaploid plants. The authors base their test on comparing the average distances between diploid plants and between randomly chosen plants. This seems to discard a lot of information. A better test would surely be whether nearest neighbors tend to be of the same or different forms?
>
> *This is a good suggestion. A test based on nearest neighbors arrives at the same conclusion as the average-distance test, and we now report both in Table 2.*

If you are rebutting a comment, then your Response must explain clearly why you believe the reviewer to be wrong, and how you've changed the manuscript so that other readers won't make the same mistake. Even if you're pretty sure the reviewer *shouldn't* have gotten it wrong, this is not the time to preserve your authorial dignity. Instead, admit—even exaggerate a little—your error in allowing the reviewer's mistake. This polishes your image of gracious cooperation, and also helps shape the way the editor thinks about your revision: you may not be doing what the reviewer wanted, but nonetheless you're using the review to improve your manuscript. For example:

> The results in Table 1 seem incorrect. The authors present a 3×2 contingency table (plants diploid, tetraploid, or hexaploid × herbivore-attacked vs. unattacked), but according to their Methods they don't actually have data on unattacked plants. So how do they get cell counts for that category? If they come from previous work, differences between sites will be confounded with attack status. Or if they come from Monte Carlo simulations, the two columns aren't really comparable. This analysis should be dropped.

---

[2] In particular, *do not* write the snappy comebacks that might tempt you. When I was a graduate student, I got a review including the comment "this part of the manuscript seems sloppily written." I found a minor misstatement by the reviewer, and wrote a Response including the zinger "the reviewer accuses me of sloppy writing, but I might say the same of the review." I felt very clever about this at the time. I now realize that feeling clever is often a strong indication that you aren't, and that reviewers, not just editors, read the Response. Anyway, the reviewer was right.

> *While the reviewer is mistaken here, we appreciate the comment because our careless writing led to the misunderstanding. The columns in Table 1 are not "attacked" vs. "unattacked," but rather "attacked" vs. "all available plants". As a result, the confounding issues raised by the reviewer do not apply. We had the column titles right, but our coverage in the text was confusing. We've clarified our wording in the Methods (111–118), Results (152), and Discussion (167–178). There is now a consistent focus on the attacked vs. available comparison and a clear explanation of why we took that approach.*

In this example I've referred to new text by line numbers but not actually quoted it. There are two schools of thought on this: some editors prefer to see the new text right in the Response rather than having to flip back and forth between Response and manuscript. I prefer the line number system because it keeps the length of the Response reasonable, and because as an editor I'm going to look to the manuscript anyway to see the changes in context.

Justification is also in order when you're only partly adopting a reviewer's suggestion. For instance (and note the offer to defer to the editors):

> The authors use both AMOVA and an NJ tree to ask whether hexaploid plants cluster together. I'm not sure AMOVA is the best tool for this; I would suggest adding a STRUCTURE analysis. I don't think the NJ tree adds anything; it should be deleted.

> *A STRUCTURE analysis is an excellent suggestion. We now provide one (Table 4), and it agrees with the other analyses (222). We see the three analyses (AMOVA, NJ tree, STRUCTURE) as complementary, in part because they make different assumptions about the biology behind the data (134–137). We would prefer to keep the NJ tree, because it provides the most visual expression of the structure in our data, and this can help the reader's intuitive grasp of what's going on. However, we realize this means a slightly longer manuscript, and if the editors strongly favor relegating one analysis to an appendix, we can do so.*

When writing your Response, you need to decide what material belongs there for the editor to read, and what belongs in the manuscript for your readers. Substantive points of science normally belong in the manuscript, not the Response, because if the editor needs them to accept the correctness of your approach, readers will need them, too. The Response should summarize the science issues, of course, but with an emphasis on why you took one approach rather than another, and how your revision differs from your original manuscript as a result.

Now you have a complete draft revision and Response, but as for anything else you write, don't immediately submit it. Instead, set it aside for a day or two and then subject it to self-revision (chapter 21). Pay special attention to tone: if you find hints of annoyance with the reviews peeking through, edit them out. It's perfectly normal to feel such annoyance (justified or not), but it's counterproductive to let it show. Finally, finish things off by finalizing the line-number references in your Response.

Now you're ready to resubmit. Of course, this may not be the end of the story—before final acceptance, some manuscripts cycle through several iterations of review, revision, response, and further review. Nonetheless, you've completed a major step. Before you move on to your next writing project, reward yourself or take a short breather. You've earned it.

Chapter Summary

- Each manuscript sent back to a journal after review should be carefully revised and accompanied by a "Response to Reviews."
- You do not have to make every change reviewers suggest, because reviewers are not always right. However, it is wise to assume they are *probably* right.
- When you disagree with a reviewer's comment, you have three (satisfactory) options: provide a rebuttal without revision, provide a rebuttal but also revise, or defer to the editor. The second option should be by far the most common.
- The "Response to Reviews" should make it easy for editors and reviewers to see that you have made appropriate revisions to satisfy their concerns.

PART VI

||||||||||||||||||||||||

Some Loose Threads

Scientific writing is a complex topic, and a few important dimensions resisted tidy inclusion in the first five Parts of this book. The three chapters of Part VI take up three loose threads. First, although most discussions of scientific writing emphasize the journal paper, over your career you'll deal with many additional forms (theses, grant proposals, peer reviews, and more)—each with its own audiences, functions, and conventions. Second, I've made little reference so far to collaborative writing, and because an increasingly large majority of the scientific literature is coauthored, it's important to think about ways to manage coauthorship so that it's a positive force rather than a source of conflict. Finally, not all scientific writers are native English speakers, and because our literature is dominated by English, non-native speakers are likely to face special challenges. Native English speakers are well advised to think about these challenges too, as nearly all of us will supervise, collaborate with, or review or edit the work of a non-native speaker.

TWENTY-FIVE

||||||||||||||||||||||||||||||||||||||||||||||||||||||||||||||||||||||||||||||||||||||||||||||||||||||||||||||||||||||||||||||||||||||||||

The Diversity of Writing Forms

When the topic of scientific writing comes up, most scientists think first of the peer-reviewed journal paper. This is typically the medium by which we communicate new results and the yardstick by which we measure our careers. It's also the form we emulate as undergraduates writing lab reports, and on which we cut our writing teeth as graduate students. The bulk of this book centers on journal papers because thinking about them is both natural and important. However, journal papers are far from the only thing that scientists write.

In chapter 6, I described my typical year's writing output. There I wanted to make a point about *quantity* of writing. Here we turn to what it suggests about *diversity* of writing. My typical-year list included journal papers, book chapters, grant proposals, peer reviews, technical reports, and administrative documents. I didn't include lectures and talks, although perhaps I should have, and in the last two years, I've also written a book (this one). Your list will differ, depending on your discipline, career path and stage, and preferences, but it will include writing of many forms. Fortunately, moving among writing forms isn't as hard as it sounds. It requires adjustments, but these flow quite simply from a bit of thought about who your readers are.

Consider Your Audience

Before beginning any piece of writing, ask yourself three essential questions:

- Who is the audience for what I'm about to write?
- Why will they read what I've written?
- What do I want them to get out of it?

The first question identifies the readers you're addressing when you write. What kind of people are they, and what do they bring to your effort to communicate with them? Most obviously, readers vary in scientific knowledge. If you're writing a journal paper, you can assume that your readers are familiar with standard theory, experimental designs, and terminology in your discipline. If you're writing an op-ed for your local newspaper, though, many of your readers won't have taken high-school science (or may have repressed the memory). Readers also differ in their situations and mental states as they pick up your work. They may read and reread your paper when they can give it their full attention; or they may skim your grant proposal among dozens of others on a tight deadline or visit your poster presentation when exhausted on the last day of a conference.

The second question identifies the reader's approach to your writing. Someone may read your writing to learn the new science you've discovered (journal paper), to be persuaded of the quality of your work (grant proposal), or to be entertained (lay essay). Your writing will be most effective when you understand why your reader has picked up your work, and deliver what they're looking for.

The third question identifies what you want your reader to learn from you. All writing aims to change the reader's mental state in some way—adding new information, changing an opinion, or whatever. If you can clearly identify your objective for the reader, you can shape what you write to reach that objective. This is closely related, of course, to the issue of finding your story (chapter 7).

Thinking about these three questions makes it clear why writing or graphics designed for one audience can rarely, if ever, be used unaltered for another. Specific audiences need specific things. All else follows from this—as I'll illustrate for a few of the non-journal writing forms you're likely to tackle over your career.

Book Chapters

Edited books are rare in some fields, such as mathematics and physical chemistry, but quite common in others, such as ecology and earth science. Such books assemble chapters by different authors to examine aspects of a broader topic. Authors are normally invited to contribute by the book's editor, manuscripts may or may not experience peer review, and rejection is uncommon.

Some book chapters report results of primary research. These needn't differ much from journal papers except that they are often longer and more detailed. Most book chapters, however, belong to the secondary literature. Some are essentially review papers. Others serve as retrospectives, synthesizing data from many published papers together with unpublished data. Still others are more like encyclopedia entries, summarizing knowledge of a topic, compiling a set of available techniques, or otherwise providing reference material (for instance, the *Encyclopedia of Earth Sciences* series; Finckl 1968–2016).

Book chapters offer more freedom of structure than journal papers. Rather than adhering to IMRaD structure (chapter 8), chapters are often more loosely narrative. This is partly because they're often broader in content than a journal paper. Departure from the IMRaD convention may seem freeing to the writer, but without careful handling it can leave the reader adrift in a wandering text. For this reason, rigorous outlining is especially important to writers of book chapters. Well-organized section heads that establish and communicate logical structure are extremely valuable to their readers.

A final distinction between journal papers and book chapters is that papers always stand alone, but in a book, each chapter relates in some way to the others. The strength of that relationship varies. Some books have strongly interdependent chapters, with each one citing others and later chapters depending on material developed in earlier ones. Contributors to such books must exchange outlines and drafts so they can write their chapters to fit together. More frequently, contributors are merely asked to exchange nearly complete drafts and to insert references to other chapters at this late writing stage; or the editors may simply distribute a list of contributors and chapter titles, and hope for some coherency.

Recent changes in the way readers access book chapters are eroding the distinction between chapters and journal papers. Even five years ago, online indexing and distribution of edited books was poor, so most readers located a book chapter in a physical copy of the whole book. This made it natural to read multiple chapters and to take advantage of integration between them. As readers increasingly access chapters digitally and separately, more editors will probably allow—and perhaps ask—chapter authors to emulate the standalone quality of the journal article.

Monographs

Scientific monographs (book-length technical pieces) were once a standard venue for primary research results—think, for example, of Newton's *Principia Mathematica* or Darwin's *Origin of Species*. With rare exceptions, monographs no longer play this role in the natural sciences. Instead, nearly all modern monographs belong to the secondary literature.

Monographs largely share function and audience with their corresponding shorter forms (journal papers in the primary literature; book chapters in the secondary literature). They present special challenges for the writer largely as a result of their length. It can be difficult to sustain a coherent logical thread in a long book with a complex argument, making outlining and chapter organization critical. Because writing a monograph takes months or even years, the project's scope, perspective, and organization are likely to shift more than once, and the first material written may need substantial overhaul once a complete draft is available.

Writers of monographs typically submit a proposal to a publisher, or a publisher may invite them to submit one. The proposal is usually considered by an "acquisitions editor" who solicits opinions of peer reviewers and an editorial board. If the proposed monograph is deemed suitable for publication, the writer may be offered a book contract specifying delivery deadlines, length, content, royalties, rights for reprinting and revision, and so on. All these things are, in principle, negotiable between writer and publisher.

The form of the proposal varies enormously among fields and publishers. It might be an outline accompanied by a sample chapter or two, or it might consist of most or all of a completed draft. Here, the interests of writers and publishers are not in alignment. The writer will prefer the briefest possible proposal, because it would be foolish to invest enormous amounts of time writing an entire monograph without knowing that the result will be published. The publisher, by contrast, will prefer as comprehensive a proposal as possible (even a complete draft), not wanting to commit to a book they haven't seen. There is no easy way out of this conflict, especially for a first-time monograph writer for whom both writer's and publisher's preferences will be most acute. If you're proposing a monograph, therefore, you should make early overtures to one or more publishers to discuss the content and form of a book proposal. It is, by the way, entirely legitimate to submit the proposal to more than one publisher at once as long as you let each publisher know you are doing so. (Such multiple submission is not allowed for journal papers.)

Technical Reports

Technical reports take many forms. You may report results of funded research to a granting agency, or summarize for a legislative body the scientific background to an issue of concern. The report may be incidental to your research (for instance, when it's a condition of receiving a permit for research in a national park) or its main product (as when you're commissioned to write a report on a topic you wouldn't otherwise have tackled). There are so many possibilities, in fact, that the only general guidance possible is to remind you of two central questions: who will read your report, and why? If you're not sure, ask. There's little point working for months to write a fifty-page, densely technical report if it will only be skimmed by an intern looking for a sentence to use on a web page.

Oral and Poster Presentations

Oral and poster presentations share with other writing forms the basic requirements of clear communication and a well-defined story. Otherwise, though, they differ dramatically, and over your career you will mutter and squirm through dozens of wretched presentations ruined by presenters who haven't thought about those differences. Detailed and excellent books on the topic are available (e.g., Anholt 1994, Davis 2005); here, I'll mention just a few distinctions.

Some differences relate to the mode of delivery. Presentations emphasize visuals, with minimal text in a supporting role. This is the reverse of other writing forms. Also, at least for oral presentations, the audience is locked into your rate of delivery, and someone who loses the thread can't flip the page back to pick it up again.

There are even more critical differences relating to the audience's mental state. You can assume readers of your journal paper are intrinsically interested in your topic. You can't make that assumption when you give a departmental seminar, to which some people come only because they come every week. At a conference, you'll speak to people who are tired (with yours the twelfth talk they've heard since dragging themselves out of bed hung over) and distracted (sneaking glances at the program as you talk to see where they're supposed to be next). Distraction is even worse at poster sessions, where someone "reading" your poster is also juggling a drink and a plate of snacks and jostling for position in a crowded room while watching for friends passing by. If all this sounds horrible, keep in mind a more positive difference: unlike the readership of a journal paper, your presentation audience is *there*. You can address them directly, see their puzzled expressions if you've left them behind, ask them questions and hear their replies, and have them contribute suggestions for your work.

Grant Proposals

We all dream of having our research funded by a wealthy patron with no questions asked, but in practice every scientist needs to write grant proposals. These vary tremendously. You may ask for a few hundred dollars or many millions. You may propose research on your own, with a few collaborators, or in a consortium of hundreds of scientists across dozens of institutions. You may write a proposal to a government, private foundation, university office, or corporation. You may be asked to make your case in a single paragraph or in a hundred pages of detail. Fortunately, most granting agencies provide detailed advice to help you meet their expectations. Read it carefully. Book-length guides to proposal-writing give far more detail than can appear here (e.g., Friedland and Folt 2009, Oruç 2012).

Some underlying principles apply to all proposals. Let's begin with the audience, which is normally a small grant panel (committee). That panel's scope might be narrow (say, molecular oncology) or very broad (earth sciences, or even all sciences). Regardless, you can safely assume that none of its members will be familiar with, or intrinsically interested in, your particular research area; instead, they have to read whatever proposals are submitted. This puts a premium on your ability to make the substance and importance of the proposed research obvious to the non-expert. While some agencies send proposals out for peer review, this doesn't change the situation much. The peer reviewers may know your research area intimately, but they only provide comments; panel members still read your proposal and decide its fate.

The function of a grant proposal is persuasion: the members of the grant panel must decide whether to fund you, and you want to explain to them why they should. Therefore, your proposal must clearly identify the research question that you propose to tackle, and then convince its readers of three things:

- **That the question is important**. Granting agencies have limited funds and can't fund all (or even most) proposals. They will prefer to fund important research—but what does "important" mean? Critically, it means important to *their* goals, not yours. If you seek funding from a chemical company, it doesn't matter if your idea puts you in line for a Nobel Prize; instead, you need to explain how your research will pay off with new or improved products or processes. On the other hand, if your proposal is for a basic-science program at the US National Science Foundation, you're wasting your time presenting a business plan to monetize your results; instead, you need to explain how your research will solve a fundamental problem in the field.
- **That the question can be answered**. In the most straightforward part of a proposal, you will pose one or a few specific hypotheses derived from your more general research question, and then explain how your methods will produce data that can test each one. This will likely include experimental designs, observational schemes, and/or theoretical approaches, along with methods for statistical analyses and their

interpretation. The amount of methodological detail required, or permitted by length, will vary enormously among agencies.

- **That you can do the work to answer it.** It's not enough to propose an important research question that can be answered; you must also establish that it can be answered *by you*. This means addressing your experience with the system in question and your expertise with your proposed methods. It means demonstrating your success in previous research (especially research funded by the agency you're applying to). It means describing the facilities and equipment to which you have access and confirming that you have any necessary permits to work in the field, certification to work with dangerous materials, and so on. One of the best ways to demonstrate capability is to provide pilot data: small datasets from work you've already done. If you can obtain and analyze pilot data, you can presumably execute the full-scale work for which you seek funding.

You may chafe at the idiosyncratic requirements of different granting agencies, the need to demonstrate capabilities that ought to go without saying, and having to spend months crafting a detailed proposal to an agency that, more than likely, will decline funding anyway. In the privacy of my office, I have certainly let fly a few impolite words on the subject. However, your first experience serving on a grant panel, faced with dozens of worthy proposals and no way to fund more than a few, will go a long way toward demonstrating the importance of getting your own proposals just right.

Theses/Dissertations

A thesis or a dissertation is the document written and defended by a graduate student, describing the research done to earn the graduate degree. In some countries and fields there are distinctions between the terms, but these are too idiosyncratic for them to carry separate meaning on global or interdisciplinary scales. For simplicity, I'll refer to all such documents as "theses."

Theses differ from most other writing forms in that they don't have a single audience or a single function. Instead, thesis writing serves as many as three distinct functions, for three distinct audiences:

- **Communicating science**. The core of a thesis is a novel contribution to scientific knowledge. The document must communicate this contribution to other scientists in your field (the audience for this function), just as a journal paper would. Well, sort of. Actually, it's uncommon for a scientist to read your results in a thesis—normally, they'll read about your work via its separate publication in the scientific literature. But it's still essential that the thesis *could* perform this function; it won't be accepted otherwise.

- **Establishing credentials**. A thesis is evaluated as evidence that its author merits credentialing in the field. The examining committee assesses the author's understanding of the methods, results, and implications of the work, and their awareness of the work's context in the field. This usually requires detailed presentation of methods and analyses, a comprehensive review of past literature, and a thorough discussion of what the results suggest for the field as a whole. All of these are normal elements of journal papers, but the credentialing function may require much more detail than is appropriate for journal publication. The audience for this credentialing function is the examining committee, who vote on whether to accept the thesis.

- **Archiving unpublished material**. In many laboratories theses serve as archives, recording unpublished material for an audience consisting of future researchers in the lab. The material recorded might include highly detailed methodologies and annotated datasets. It might include records of failed experiments as well as successful ones (so future grad students don't run down the same blind alleys). It might include data and analyses that weren't included in manuscripts for publication because they didn't fit into the stories those manuscripts told. The degree of detail desired here may far exceed that needed for the communication and credentialing functions.

Expectations for these three functions vary considerably among theses. For course-based degrees, publication in the scientific literature may not be expected, and the credentialing function will be emphasized over the other two. For research-based degrees, the weighting of the three functions varies among institutions, programs, and examining committees. It is wise, therefore, to explore expectations with your committee before devoting effort to thesis writing.

The need for three audiences and three functions to coexist in a single document explains many features of thesis structure, including the existence of two quite different standard formats: the "thesis" and "papers" formats. Most programs offer both options. A thesis-format thesis consists of a single lengthy document: it may be divided into chapters, but these aren't intended to stand on their own. It will often contain much more detail than could be published in a set of journal papers. The thesis format emphasizes the credentialing and archiving functions, with the communication function implicit. (The thesis won't be accepted if it doesn't report good science clearly, but the actual communication will happen later when condensed and extensively rewritten material appears in journal papers.) A papers-format thesis, in contrast, consists of a series of chapters that are simply manuscripts prepared for journal publication, plus (usually) introductory and concluding chapters putting the work in broader context. In the papers format, the three functions are compartmentalized. The manuscript chapters address the communication function just as journal papers do, because they *are* journal papers. The introductory and concluding chapters serve only the credentialing function, and will rarely be read by anyone outside the examining committee. The archiving function is served by appendices, which can be easily skipped by nearly all readers.

The thesis format is older, but the papers format is increasingly the default. This probably serves both student and science well, as the compartmentalization allows different parts of the thesis to be tailored to their different audiences and functions. Furthermore, the papers format requires the least additional work to adapt the thesis for publication in the literature (the main route for disseminating the findings). Pockets of thesis-format fans persist, though, not just because of academic tradition but also because that format allows the student to display mastery of a field in a way not easily compatible with any other writing form.

Peer Reviews

Like theses, peer reviews of manuscripts submitted for publication have more than one audience. They are written for a handling editor and for the manuscript's author(s), and serve a different function for each. For the handling editor, your review communicates your opinion of whether the manuscript should be published, and if so, what changes should be made first. This is an evaluative function. For the author(s), in contrast, the review's function is to improve the quality of the manuscript.

Some journals separate the two functions of a review. For such a journal, you'll write a paragraph or two recommending and justifying an editorial decision (chapter 23), addressed to the editor and not copied to the author(s). A separate and longer document, addressed to the authors, carries your specific criticisms and suggestions for improvement. Other journals combine the two functions in a single document, nominally addressed to the editor but in fact written for and seen by both audiences. When the two functions are separate, you can be a bit blunter with your recommendation, but otherwise there isn't a lot of difference.

What you should actually *say* when you write a review is one of the deeper mysteries faced by beginning scientific writers. Except for grant proposals, all the writing forms discussed here are by design public documents: it's easy to find excellent models. Sample grant proposals are available from some (but not all) agencies, while colleagues are usually happy to share grant proposals they've written. But when you're first asked to write a review, you may have seen only the few you've received in connection with your own submissions. You can't ask colleagues to show you reviews they've written, because it's unethical for a reviewer to share the reviewed manuscript with anyone else, and that makes it arguably unethical (and not very useful) to share the review itself. You can, however, ask colleagues to share reviews they've *received*.

The best reviews are dispassionate in tone, being critical of the work when necessary, but never of its author. They are careful to serve both functions, and to distinguish them (for example, communicating whether a criticism should preclude publication or is merely an avenue for improvement). They are detailed, referring to problem areas in the manuscript by line number or at least by section. They provide concrete criticisms and suggestions. For example, a review that says "The statistical analysis is wrong" is of little value. A good review will say "The statistical test at line 377 appears problematic because the assumption of normality of residuals is likely violated for this kind of data." An excellent review will add "(see treatment by †Diallo et al. 1990)" and "a model with Poisson errors or a randomization test might be an alternative." If in doubt, apply the golden rule: do unto those you review as you would have them do unto you.

A final issue that arises is the decision to remain anonymous or to sign your review. Nearly all journals offer both options. An argument for signing is that it permits an author to contact you if what you've written is unclear, or if further discussion could improve the manuscript. This certainly benefits the author; but signing can benefit the reviewer too. A constructive (even if critical) signed review can establish you in an author's mind as a future collaborator, or at least someone to whom a small debt can one day be repaid.

Consider some arguments against signing, however. Active retaliation for negative reviews is rare, but not unheard of. And signing can lead to awkward situations. Imagine, for example, that Dr. Smith's manuscript has just been rejected based on your intensely critical review. Should Dr. Smith next month happen to review *your* manuscript, consider you as a job candidate, or the like, your negative review should and probably will have no influence—but both you and Dr. Smith may prefer having no reason to suspect it might.

The decision about anonymity involves a tradeoff that may shift over your career. When the stakes are higher (for the early-career researcher, before hiring, or perhaps before tenure), it may be wisest to routinely remain anonymous. More established researchers may sign most or all of their reviews.

For some excellent step-by-step advice on writing peer reviews, see Nicholas and Gordon (2011).

Blogs

Science blogs have become quite common (mine is http://scientistsees squirrel.wordpress.com). Some focus on science outreach, presenting nontechnical pieces intended for lay readers; others include technical pieces that are essentially commentaries on, or expanded versions of, primary-literature papers; still other discuss issues relating to the culture and practice of science. These are three very different functions, with different audiences and thus different needs for writing. While in some fields (such as economics) blogs now play an important role in the development and dissemination of research results, this function appears to be very unusual for science blogs. The influence of blogs on the way scientists think about and practice science, on the other hand, seems to be strong and growing stronger.

In terms of writing, blogs differ in several ways from most of the other forms discussed here. First, because a blog is online, you can include hypertext links to a virtually unlimited array of other online resources. This is both an opportunity (to provide context, background, supporting material, and arcane but amusing connections) and a risk (of losing a reader who clicks away and never comes back). Second, online writing can be very informal. Even blogs with quite technical content can take a conversational style and include colloquialisms, grammatically nonstandard writing, witty asides, and even emoji. Third, and perhaps most substantively, blogs allow multidirectional communication: readers can leave "Replies" to your post or to each other, and you can respond. This can be wonderful, when readers add perspective or content you hadn't thought of. It can also be annoying, because inevitably, someone will leave the kind of comment the internet is famous for: *ad hominem* attacks, sexist or racist jokes, spam, or just irrelevance. If you write a blog and enable Replies, you should have a policy on what kinds of reader responses you'll allow, and which you'll block.

Other Forms

This survey is far from complete. I haven't covered, for example, textbooks, administrative reports, or lay essays. Exhaustive treatment of every writing form would make this book a doorstop. But the key to every writing form is the same: ask yourself who will read what you write, why they will read it, and what you want them to take from the experience. Once you know the answer to these questions, you know what your reader needs and you can craft your prose to supply it with telepathic clarity.

Chapter Summary

- Scientific writers produce many forms in addition to the journal paper: for example, book chapters, monographs, technical reports, oral and poster presentations, grant proposals, theses, and peer reviews.
- Different forms have different audiences and different conventions for style and content.
- Writing for different audiences means asking yourself three questions: Who is the audience? Why will they read what you're writing? What do you want them to get out of it?
- Book chapters and monographs are unusual for early-career writers in most scientific fields.
- Presentations must be very different from material intended for reading, because audiences are likely to be broader and less committed.
- Grant proposals must persuade agencies of three things: that your research question is important, that it can be answered, and that you are capable of doing the work to answer it.
- Theses serve three functions: communicating science, establishing credentials, and archiving material that won't be separately published. The latter two functions explain the differences between theses and sets of published papers.
- Peer reviews should be dispassionate, criticizing the work but not the author. They should provide concrete criticisms and suggestions.
- Blogs are relatively informal, and allow back-and-forth exchange with readers. Their role varies across fields and may evolve rapidly.

TWENTY-SIX

Managing Coauthorships

A peculiarity of the preceding chapters is that they discuss writing as something done by a writer. At first glance that seems completely unremarkable, even tautological: who *could* possibly write, other than a writer? The answer, of course, is *a team of writers*. Through your career, you'll probably write with one or more coauthors far more often than you write alone. This is especially true for journal papers, so I'll focus on those here, but the issues are universal.

In some ways coauthorship doesn't matter to writing: a dangling participle obscures meaning no matter how many authors conspired to commit that particular sin. But in other ways, coauthorship matters very much. Having someone else to share your writing process can solve some writing challenges. It also presents challenges of its own.

Coauthorship Past and Present

Francis Bacon's vision of science in *New Atlantis* (Bacon 1627) was radical for its time. In chapter 1, I emphasized Bacon's notion of scientists *communicating* with each other as a departure from the secrecy of medieval science. Bacon also described scientists *collaborating* with each other—a bizarre idea to medieval scientists, who were generally solitary if not reclusive.

Bacon's idea of clear scientific communication caught on fairly rapidly, but collaboration and coauthorship remained rare for a long time. From 1665 to 1800, collaborative or coauthored papers account for a tiny fraction of published papers (Beaver and Rosen 1978): from less than one percent in biology to just under five percent in astronomy[1]. Coauthorship rates grew slowly through the nineteenth century before beginning to climb more steeply in the early twentieth century. By 1980, seventy-five percent of journal papers across the sciences were coauthored, and by 2000, the fraction had reached ninety percent (Glänzel and Schubert 2005). Even in mathematics, long the most individual of scientific fields, coauthorship rates rose from about thirty-three percent to sixty percent between 1980 and 2000 (Glänzel and Schubert 2005), and by 2012 it had reached seventy percent (King 2013). The average number of authors per paper has grown too, and papers with dozens or even hundreds of authors are no longer rare. The coauthorship record seems firmly in the grasp of experimental particle physics, largely because of an authorship convention held by several research groups in which every scientist or engineer working on or with a detecting instrument is an author on every paper published during their tenure with the group. These groups are huge, and in 2015 two of them combined for a 5,154-author paper (Aad et al. 2015) in which the author list alone takes up almost twenty-five of the paper's thirty-three journal pages. Perhaps fortunately, this authorship practice remains rare.

Early growth in collaboration was probably spurred by the professionalization of science, which increasingly had scientists working in institutes and universities alongside their colleagues (Beaver and Rosen 1978). Rapid growth in the twentieth century was fueled in part by funding opportunities for "big science" that required collaboration, and facilitated by technological advances that made it easier for scientists to travel and to interact at a distance (Glänzel and Schubert 2005). However, practices that first arise for identifiable external reasons often evolve into cultural norms and become self-reinforcing. In the sciences, coauthorship is now something of a default way of working and writing.

---

[1] Low as they are, these figures are probably overestimates. In the seventeenth and eighteenth centuries, authorship conventions hadn't yet matured. The first "coauthored" contribution I can find is the wonderfully titled "An Extract of a Letter Containing Some Observations, Made in the Ordering of Silk-Worms, Communicated by That Known Vertuoso, Mr. Dudley Palmer, from the Ingenuous Mr. Edward Digges" (Palmer and Digges 1665), but in fact Palmer's role was only, as a member of the Royal Society, to forward to the Society a letter from his cousin Digges, a nonmember. Many early "coauthored" publications were similar, and they wouldn't merit coauthorship today.

In fact, a *curriculum vitae* listing only single-authored papers would be highly unusual in most fields, and might even provoke suspicion about its subject's collegiality. All authors, therefore, need to know how to manage coauthorships.

Who Is a Coauthor?

The coauthorship issue over which the most ink (and anguish) has been spilled is one that should arise long before writing actually begins: who will be a coauthor on a particular publication. Someone who lends equipment clearly merits only a mention in the Acknowledgements, while someone who's an equal intellectual partner at all stages of research is clearly a coauthor. But between these simple cases lies a long continuum in which decisions about coauthorship are not obvious—or worse, where different decisions seem equally obvious to different participants. As a result, serious disputes over authorship are not uncommon and occasionally end up in legal action or lifelong enmity. The good news, however, is that coauthorship disputes are entirely avoidable, if you do two things: first, take advantage of published guidelines for determining coauthorship, and second, discuss authorship openly and honestly with collaborators before beginning work. *You have to do both.*

Guidelines for coauthorship are easy to find, because they are offered by journals, scientific societies, granting organizations, universities, departments, and even individual research laboratories. A set from the International Committee of Medical Journal Editors (http://www.icmje .org/recommendations/browse/roles-and-responsibilities/defining-the -role-of-authors-and-contributors.html) is fairly typical:

- "Authorship credit should be based on 1) substantial contributions to conception and design, acquisition of data, or analysis and interpretation of data; 2) drafting the article or revising it critically for important intellectual content; and 3) final approval of the version to be published. Authors should meet conditions 1, 2, and 3 . . .
- "Acquisition of funding, collection of data, or general supervision of the research group alone does not constitute authorship.

- "All persons designated as authors should qualify for authorship, and all those who qualify should be listed.
- "Each author should have participated sufficiently in the work to take public responsibility for appropriate portions of the content."

Notice that these guidelines envision authorship arising from *substantial* contributions to *multiple aspects* of the work. Intellectual contributions are more important than financial or administrative ones (second bullet); in particular, the practice of principal investigators taking routine authorship on every paper arising from their laboratories is discouraged. Finally, the notion of responsibility (last bullet) is interesting and important, because it implies that coauthorship carries risks as well as rewards. If work is flawed or fraudulent, every coauthor will share the stain whether or not they were aware of the problem. For this reason, you should never accept coauthorship of a paper you don't thoroughly understand.

Authorship guidelines are very helpful in setting expectations, but for several reasons guidelines alone cannot guarantee peace. First, their very ubiquity means that it's never automatically clear which guidelines might apply—for example, is my latest collaboration subject to guidelines adopted by my university, my collaborator's university, one of the several societies to which we both belong, or the journal in which we propose to publish? Second, coauthorship practices vary among countries and fields; for instance, in North America solo publication by Ph.D. students of their thesis chapters is rare among chemists but common among ecologists. Third, the huge variety of possible contributions means that all guidelines have to include imprecise words such as "substantial." Guidelines can help begin a conversation among collaborators about coauthorship, but they can't replace one.

The discussion about coauthorship should take place among all collaborating parties, as early as possible in the research (not writing!) process. It should produce a written agreement of who will do which pieces of the work, and what authorship will result. This seems simple and obvious, but in practice these conversations very often don't happen. Some scientists believe they don't need to discuss coauthorship because they know their collaborators well or have worked with them before without problems. (This is a bit like reasoning that you don't need to wear your seatbelt because you've never been in an accident before.) Others find initiating coauthorship conversations awkward, perhaps because they think it implies that they don't trust their collaborator(s) to do the right thing about authorship later on.

If you're in the awkwardness camp, there are two ways to help defuse the tension. First, you can make it clear that your raising the topic is spurred not by the individual collaborator, but by the general difficulty of making unambiguous rules about coauthorship. Second, you can deflect some of the responsibility by referring to what you've read about the importance of coauthorship conversations. So you might open the discussion with a gambit like one of these:

- *"As long as we're talking about who's going to do what on this project, let's talk about authorship too. I know there are lots of different ideas out there about how authorship is decided, and I'd hate to have us surprised later if we happen to have different ones."*
- *"You know, I was reading the other day about making authorship agreements early in a collaboration, and it sounded like a good general practice. Let's talk about that now and get it out of the way."*
- *"Last week my old advisor was telling me horror stories about authorship disputes, and how she now makes authorship agreements with collaborators before starting a project. That sounded like a good idea to me—should we do the same thing?"*

If a bit of awkwardness lingers, consider it an investment in avoiding the vastly greater awkwardness that comes when your friend and collaborator sees a published paper they thought you had worked on together, but without their name on it (quickly making them an ex-friend and an ex-collaborator). In the big scheme of life awkwardness, coauthorship conversations should be pretty minor stuff—they've got nothing on a high school dance!

Order of Authorship

Once a list of coauthors is set, the remaining loose end is the order in which those authors will occur in the paper's byline. If there are two coauthors, who gets to be first author and who second? If there are eight, who gets stuck in the obscure sixth position?

For indexing purposes, this doesn't matter: it's just as easy to locate papers on which Dr. Olga Kashirin is the fourth author as those on which she's first. However, authorship order matters when your curriculum vitae is evaluated, as it might be for a grant, a job, or a promotion. Usually only the first and last authorship positions really matter, with the "interior" positions lesser and interchangeable in importance.

So who should be first author, and who should be last, on your coauthored paper? There is no universal convention. In mathematics, economics, and theoretical physics, multiple authors are nearly always listed alphabetically—a system that's simple, unambiguous, and fair, but surrenders an opportunity for authorship order to convey information. Alphabetic schemes are also common for papers with very large numbers of authors. In fields where alphabetization is not conventional, there are at least three different systems. In some fields (for instance, ecology) the first author is the person who contributed most to the research and to writing the manuscript. In others, the principal investigator (graduate advisor, team leader, or senior scientist) goes first, regardless of who did the most work or writing. In still others (for instance, molecular biology), the principal investigator is listed last. If that isn't complicated enough, practices can vary within a field through time, among journals, or from country to country. In organic chemistry, for instance, North American journals once followed principal-investigator-first but now conform strictly to principal-investigator-last, while European journals may use alphabetical, principal-investigator-first, or principal-investigator-last schemes. Finally, it isn't unusual to see a footnote dispelling conventional expectation: specifying, for example, that two authors contributed equally to the work[2]. Phew! If you're not familiar with authorship-order conventions in your field, ask a more senior colleague.

---

[2] Or more interestingly: specifying that order of authorship was based on "random fluctuations in the Euro/Dollar exchange rate" (Feder and Mitchell-Olds 2003), "a flip of what William B. Swann, Jr. claimed was a fair coin" (Swann et al. 1990), "rock, paper, scissors" (Kupfer et al. 2004), or "brownie bake-off" (Young and Young 1992).

Writing with Coauthors: Logistics

Let's assume that you've agreed to cowrite a paper with one or more coauthors. (I'll draw a distinction between cowriting and coauthorship, as under some authorship practices a particular coauthor may contribute to the research but not to the writing phase of a project.) How should you go about actually writing as a team? There are probably as many answers to this as there are teams of cowriters, but here are some suggestions.

- **Together or apart?** Writers new to cowriting are sometimes surprised to find out how uncommon it is for two writers to write "together" (working on the same paper in the same room at the same time), and how ineffective it is when they try. Instead, nearly all cowriters arrange for each to work on an identified part of the manuscript, with drafts exchanged from time to time and no need to ever be in the same place. The only real exception, and the primary reason for nearly all episodes of writing "together," is that dedicating a time to having cowriters focused on the same project in the same place can serve as a commitment device to enforce both writers' focus on the project. Even if you're working without talking and with your backs to each other, at least you're conscious of your agreement to get the work done.
- **Identifying a lead writer**. The first task for a writing team is identifying a "lead writer." The lead writer needn't be the first author, and needn't do the majority of the actual writing. Instead, the lead writer shepherds the other cowriters through the writing and publishing process, retains the "definitive" version of the manuscript as changes are made by the various cowriters, and edits to ensure overall consistency of style and format. When the manuscript is submitted for publication, the lead writer will be the corresponding author, communicating with the journal (and eventually with readers) on behalf of the authorship team. Things tend to go well when the lead writer is good at keeping track of details and sticking with a schedule, and better still when the lead writer has a strong incentive to move the project

along; for this reason, early-career researchers building their CVs for an upcoming job search make ideal lead writers. Even with the best of intentions, busy academics can easily let one project slip, and a lead writer who can strike the right balance of nagging and understanding can keep a manuscript from drifting rudderless, nobody's top priority on any given day.

- **Dividing writing tasks.** One of the great advantages of cowriting is that writing tasks can be divided up in ways that take advantage of complementary strengths. Two common strategies are division by section or by writing stage. Having different cowriters draft different sections lets you exploit differences in experience: for instance, the person who ran an experiment can most easily draft the Methods, while the person who executed the statistical analysis can most easily draft the Results. The Introduction and Discussion may be best suited to the writer who knows the literature best: perhaps a junior member who has just jumped through a graduate-school hoop such as a comprehensive exam. Roughly, this divides the work in parallel, with multiple sections being drafted at once. Alternatively, tasks can be divided in series: for instance, with one writer responsible for the complete first draft and then another taking on revisions. This can capture complementary writing strengths. For example, a friend of mine excels at producing chapter first drafts quickly. However, these drafts are awful, and unfortunately my friend hates revision and takes a long time to make them any better. I, on the other hand, am glacially slow at first drafts, but once I have a complete manuscript in front of me—no matter how bad—I'm pretty good at whipping it into shape. The two of us have yet to write a paper together, but when we do, we'll be unstoppable.

Assigning tasks, though, isn't quite enough: the work also needs to be scheduled. Each cowriter needs to commit time to the project. Be realistic about your ability to complete your own portion of the work: it will only lead to bad feelings later if you breezily guarantee you can write the first draft in three weeks but then realize that the new course you're teaching means you can't even start for two months. The lead writer should assemble a work schedule, which might look something like this: "*Writer A will complete analyses and draft the Methods and Results by the end of September. By then Writer B will be back from*

Guam, and will draft the Introduction and Discussion by mid-November. I'll combine these and pass them to Writer C for a thorough revision. The revision will be sent before winter break to A and B for comments. Those comments are due in mid-January, and I'll handle final revisions and submit by the end of February." The lead writer will then track progress, laying guilt trips on other writers as needed, to keep the project on track.

Writing with Coauthors: Nuts and Bolts

Most sets of cowriters exchange drafts electronically while working at different times and in different places. There are some ways to make this exchange work more smoothly:

- **Take advantage of software designed for collaborative writing**. There are two varieties: that designed for sequential cowriting and that designed for simultaneous cowriting. Sequential cowriting tools, such as "Track Changes" in Microsoft Word, allow a writer to edit a document while displaying both the original text and the changes made. The marked-up document can be sent to a cowriter, who can see each proposed change and either accept it, reject it (returning to the original text), or offer an alternative. The advantage of sequential cowriting is that there's no risk of two writers making conflicting changes to the same text at the same time. This disadvantage is that one cowriter may be stuck waiting for the next revision from a colleague. Tools for simultaneous cowriting, such as Google Docs, are quite different in that they allow multiple writers to edit a document in real time. Nobody needs to wait for anybody else; however, it can be difficult to keep track of who made which change and why, and frustrating to work with text that keeps changing as the other writer works at the same time. Sequential cowriting is most common, but as software improves and writers become more familiar with simultaneous cowriting, the latter might suit some teams of writers well. Try it once and find out.
- **Have the lead writer maintain a version of record throughout the cowriting process**. It's frustrating for one writer to make revisions, only to discover later that another writer had already changed the

same passages. At least under sequential cowriting, the lead writer should receive revisions from each cowriter, check the document for consistency, and give the draft a new file name including its date. This can then be sent out to the next cowriter for review, and so on. Drafts can be exchanged via e-mail or uploaded to a cloud service such as DropBox or Google Drive.

- Communicate with your cowriters through comments in the developing text. As a solo writer, if you find yourself unsure of what's best at any point in the text, you have little choice but to struggle through (see chapter 6). But when writing as a team, why not call upon the resources represented by your cowriters? You can use a comment-marking tool in your word processor, or an in-text convention such as setting off comments to cowriters with asterisks. My own drafts are sprinkled with notations like "**Arne: do you know of a good citation for this?**" and "**Julia: I can't decide whether this paragraph should come first or second—please move to suit your preference**." If you have an idea for a section that isn't "yours" to write, insert suggestions in the appropriate spot, again marked as a comment. Your cowriters will nearly always find this helpful—and if they don't, well, most of us have a pretty good idea where the Delete key is.

Agreement before Submission

While most cowriting practices are flexible, one is not: the need for submission to be approved by all coauthors. Both at initial submission, and again when revisions are submitted after review, the corresponding author must indicate that all coauthors agree to submission of the work in its current form. Usually this means ticking a checkbox on an online submission form, or including a pro forma sentence in a cover letter.

Occasionally you may be tempted to tick the "all authors approve" box even though you haven't actually run the final draft by every coauthor, or even though one coauthor wants changes that nobody else agrees with. Doing so in the face of serious disagreement would be unethical, and in any case this isn't likely to work: a routine part of manuscript handling at most journals is e-mail notification of each coauthor that a manuscript has been submitted with their names on it. Should one listed author reply that you didn't have their approval, the journal will immediately return the manuscript, and the result is embarrassment for the corresponding author and considerable awkwardness for the authorship team. With large sets of coauthors it can be something of a hassle to collect active "OK to submit" responses from all parties, so the lead writer may prefer to survey coauthors with the default response being assent. A message like this will do: *"Here's the completed manuscript, which I think is ready for submission. Please let me know by (date about 2 weeks away) if you think further edits are needed; otherwise, I'll presume you approve submission."*

The need for agreement by all coauthors raises an interesting potential problem, rarely encountered but potentially career-crippling when it arises. What happens if you're ready to submit a manuscript, but one coauthor is unavailable to approve submission? If your coauthor is deceased, your submission can be accepted smoothly, but nearly any other reason for unavailability can keep a manuscript in limbo indefinitely. A coauthor might be unavailable for many reasons: a prolonged field trip in a remote area, a medical situation such as a coma[3], a stress leave from employment in which relief of all duties is part of the prescribed care, suspension from duties for misconduct, or personal issues such as bereavement. None of these is frequent; but none is unheard of either, and I've seen more than one manuscript held up for reasons on this list.

If the period of coauthor unavailability is known and is a matter of weeks, little harm is likely to be done by simply waiting for a return. However, when unavailability is open-ended or stretches beyond a month or two, the situation is more serious. Long delays can jeopardize publication priority for a result or can seriously damage the career of a junior scientist hitting the job market or coming up for tenure. About the only way to avoid this situation is with a document I call a "publishing power-of-attorney" (Box 26.1). This is a short letter specifying that, should its signatory be unavailable to assent to submission of coauthored manuscripts, another named individual is authorized to consent on their behalf. If you are preparing a publishing power-of-attorney,

---

[3] Believe it or not, a large publisher recently assured me that they would accept a submission with a coauthor who was since deceased, but not one with a coauthor who had since entered a coma (both cases, fortunately, hypothetical). The rationale was that the deceased colleague was presumably not going to come back to contest the submission, whereas the one in a coma could wake up and launch a lawsuit.

you'll want to name someone who knows your work well enough to judge the quality of a manuscript on which your name appears, someone you trust to make an authorship decision you'll be happy with when you return from your period of absence. Most likely, this will be a close colleague in your field, perhaps a frequent collaborator. You should then place a copy on file with a colleague or a department office, and routinely tell your coauthors where to find it in case of need.

Box 26.1 Sample text for a "publishing power-of-attorney"

There is no standard wording for publishing power-of-attorney documents, which are still very rare. It is best for the wording to be as simple and as clear as possible. This sample is written broadly, so it can remain on file and serve in all situations, but it's easy to adapt to restrict the document to a particular manuscript, project, or set of coauthors (such as grad students and postdocs).

[Date]

To whom it may concern:

Should I be unavailable to participate in the publication process for a period longer than 60 days, I authorize my colleague Dr. Agathe Magnussen of the University of Jotunheim to approve on my behalf submission of manuscripts coauthored by me for publication in the scientific literature. This authorization applies to any period of unavailability falling, in whole or in part, in the period January 1, 2014 – December 31, 2016. The Chair of the Department of Biology, Niflhel University, will confirm my unavailability for the purposes of this authorization.

Sincerely,

[Researcher's name]

Publishing power-of-attorney documents are not yet common, and in my experience most researchers are dismissive of the idea until the need for one has cropped up—when, of course, it's too late. I strongly recommend that you be the first in your lab or research group to write and sign one. Once you've done so, talk to your collaborators, your advisor, and your peers about why they should do the same. Sure, lightning rarely strikes, but when it does, it hurts.

Why Bother?

Perhaps this chapter has made it sound as though coauthorship is complicated and fraught with pitfalls and frustrations. It is—but so is solo authorship! Coauthorship is worth every bit of trouble it brings, because the trouble is almost always outweighed by the myriad ways it makes research and writing easier. Coauthorship can also be enormous fun. My most frequent coauthor and I have written nine papers together— among which are papers much better, and much more easily written, than I could ever have hoped for writing solo.

Chapter Summary

- Most papers in most fields are now coauthored, and coauthorship rates continue to increase.
- Criteria for authorship vary, but most suggest that authorship is merited by substantial contributions to (1) conception, experimental design, data acquisition, data analysis and/or interpretation, and (2) drafting of the manuscript.
- Negotiation of coauthorship should occur as early as possible in a collaboration.
- There are many models for coauthorship, but it's always a good idea to identify a "lead writer" to shepherd the rest through the process.
- All coauthors need to approve submission of manuscripts. A "publication power of attorney" document can prevent publishing delays if a coauthor becomes unavailable.

TWENTY-SEVEN

‖‖

Writing in English for Non-Native Speakers

Scientific writing is difficult for everyone (chapter 2), but you may face special challenges if your native language is not English (which makes you an "EAL" writer, for English as an Additional Language). That's because the vast majority of the scientific literature— since the beginning of the twenty-first century, over ninety percent—is published in English (Ammon 2012). An even greater fraction of *read* and *cited* literature is in English because, among other reasons, citation databases emphasize English-language journals, and these journals have broader distribution and higher impact than those in other languages (Hanauer and Englander 2013).

Science has nearly always had a dominant language (or a few of them), although it hasn't always been English. Throughout the history of science, just seventeen languages have accounted for the vast majority of published scholarship, including ancient Greek, Latin, Arabic, Chinese, Japanese, Russian, and several western European languages (Gordin 2015). Latin had a long period of dominance (twelfth through seventeenth centuries). This was followed by several centuries in which a handful of languages fought it out: for instance, from 1880 to 1910, the literature was split nearly equally among French, English, and German, and as recently as the early 1970s, over twenty percent of the world literature was in Russian (Ammon 2012). But English built steadily through the twentieth century, to the point where Gordin (2015) describes modern science as "the most resolutely monoglot international community the world has ever seen" (p. 2). The reasons behind this are complex, but the result is simple: even if English isn't your first language, you'll almost certainly want to publish in it anyway. (While research of local and applied interest is more frequently published in non-English journals, few scientists will publish only such work.)

You can certainly see the dominance of English as a social-justice issue, putting scientists from the non-English-speaking world at a disadvantage. That's a topic for another book (Ammon 2012, Hanauer and Englander 2013). In this chapter, I identify potential writing challenges of interest to EAL writers, and provide some relevant advice. While I address my remarks to the EAL writer, if you're a native English speaker I hope you'll read on, because scientists who supervise, collaborate with, or review or edit the work of EAL writers should also think about the challenges those writers may be working to overcome.

Most scholarship on EAL scientific writing has focused on identifying perceived barriers to publishing in English and suggesting institutional correctives to these barriers. Few studies have instead tried to identify steps individual EAL writers might take to improve their scientific writing. Here I provide the most concrete advice I can. EAL writers are not a homogeneous group, however. They speak different native languages, have encountered English at different ages and in different educational contexts, and may be working in countries that primarily speak their native language, English, or another language altogether. Therefore, no single piece of advice will apply to all.

To Translate or Not?

One obvious question is at which stage of writing an EAL writer should shift into English. There are at least three choices: write a complete paper in your native language, then translate it into English; outline in your native language but write the text in English; or write in English from the earliest stages.

Writing in your native language can be appealing because it's easier, allows more sophisticated expression (and thus thinking) during the writing, and permits direct reuse of native-language material that might be available from presentations, lab notebooks, and so on. This strategy appears to be fairly common among EAL writers, especially early in their careers (Gosden 1996, Hanauer and Englander 2013). However, writers who adopt the write-then-translate strategy often report dissatisfaction with the results, for two major reasons. First, translating your own work takes tremendous time and effort, while those who pay for professional translation often discover that translators' unfamiliarity with science and scientific writing is a significant hindrance (Flowerdew 1999). Second, the retention of non-English phrasing and structure can mean the translated product is of poor quality. Writers of different linguistic backgrounds might use different ways of organizing material and building arguments (Mok 1993, Khatib and Moradian 2011), the translation of which can produce English text that seems disorganized. If you write first in your native language, therefore, try to think of the work as something to be written over again in English rather than something to be translated.

I'm aware of no data bearing on whether an EAL writer should outline in the native language or in English, although I suspect either can be effective. However, if you outline in your native language, it's wise to switch into English before taking things as far as a topic-sentence outline (chapter 7). Organization at the paragraphs-to-subsections scale may be influenced by linguistic conventions that differ between English and your native language.

Ultimately, it may seem almost uncaring to recommend that you write in English rather than in your native language—something that, inevitably, is difficult. However, what literature exists suggests that the English-first strategy is (overall) not much harder than the write-then-translate alternative, and it's likely to produce better text. It also offers the most potential for long-term learning.

Learning

Learning an additional language is, of course, a substantial task. For an EAL scientific writer, that task has two overlapping components: learning general English writing, and learning scientific English writing. For most EAL writers, the toughest challenges are the general-English ones of vocabulary and sentence construction (Cho 2009). However, some issues particular to scientific English can present problems: for example, appropriate use of hedging (see chapter 13) and the phrasing of criticism.

Approaches to improving one's general command of English are beyond the scope of this book, but a few strategies with special relevance to scientific writing include the following:

- **Read extensively from the English (scientific) literature.** In chapter 2, I argued that even native speakers should read, and read with conscious attention to aspects of writing they might later model or avoid. If you're an EAL writer this is especially crucial, because you can use the opportunity to notice differences in expression or structure between English and your native language. Given the dominance of English in the scientific literature, you will have to read extensively in English anyway, simply to master the knowledge base of your discipline.

- **Practice.** Frequent and sustained writing practice will benefit all writers, but EAL writers may find it especially valuable. If you are part of a cowriting team (chapter 26), volunteer to write early drafts so you can learn from revision by cowriters with more English experience. Any practice can help, though, not just practice at *scientific* writing: you may benefit from using English in blogging, e-mail, peer review, and any other opportunity you can find to practice your craft.

- **Work at an English-language institution.** EAL scientists who have spent time at an English-language institution (as a graduate student, postdoc, sabbatical visitor, etc.) widely acknowledge its value (Flowerdew 1999, Cho 2009). Given the success of immersion-style language teaching in general, this is not surprising. Attending such an institution may help most directly with your comprehension and spoken English, but it also offers opportunities to practice writing (in coursework or theses, or in cowriting with native-English collaborators) and to review writing by native-English scientists. Of course, this option may not be open to every EAL writer, because of financial constraints or reluctance or inability to leave the home country. Nonetheless, if you're able to travel, there is probably no more effective way to boost English-language ability.

- **Write the thesis, or present research, in English.** Many EAL graduate students, even if they attend native-language graduate programs, will have an option to write the thesis in English. (This is common, for example, in Quebec, Scandinavia, and China.) Granted, doing so may extend the time needed to write the thesis, but this should pay off in reduced effort to publish the thesis chapters—and keep paying off in increased English writing fluency. Similarly, conferences sometimes offer opportunities to present in the native language or in English; the former will be easier, but the latter more useful in the long run.

- **Model available written English**. Many EAL writers report choosing an existing paper related to their own work and then modeling their own writing on its structure and phrasing. This is a logical extension of the "reading" strategy (and it need not be restricted to EAL writers; chapter 5). The risk associated with this strategy is that too-close modeling can result in plagiarism. It's safer to imitate a paper's large-scale organization and its finest-scale phrasing than to model sentence- and paragraph-length pieces of text. It's also safer to model a paper that's not too close in subject, scope, or approach to your own manuscript, lest you be tempted to simply graft your own data into someone else's writing.

- **Participate in specialized writing workshops and courses.** General EAL writing courses are widely available, but occasionally you may have access to workshops or courses specifically intended to help with scientific writing. For example, Texas A&M University has offered an "Intensive Course in Research Writing" that was open to all, but specifically welcomed EAL writers. Such resources are most likely to be available at large research universities. The advent of massive online open courses (MOOCs) presents an interesting avenue for wider access, although it's not yet clear how effective online courses will be for teaching writing.

Knowing What You Don't Know

Different EAL writers struggle with different aspects of English, and just like native-English writers (chapter 21), will benefit from awareness of their particular writing weaknesses. Some of this variation is individual, of course, but quite a bit arises from the way *other* languages differ in grammar, structure, and organization from English, as well as from cultural variation that's correlated with language. If you're an EAL writer, being aware of the issues that trip up speakers of your native language is important for improving writing and can also help you understand the reactions native-English reviewers have to your work. A few common trouble spots include:

- **Articles**. Many languages lack articles (*a, an, the*), including Chinese, Korean, Japanese, Russian, Persian, and Finnish. The conventional use of articles in English is not straightforward (DiYanni and Hoy 2001, their chapter 17), and can be a major challenge for writers from these language backgrounds.
- **Countable nouns, noncountable nouns, and plurals**. A countable noun has singular and plural forms (*an apple, two apples*), while a noncountable noun is an undifferentiated collective (*the equipment, more equipment*). Not all languages assign a noun to the same category: for instance, *equipment* is countable in French (*un équipement, des équipements)* and German (*die Ausrüstung, die Ausrüstungen*). Other languages handle plurals quite differently from English: for example, most Chinese and Japanese nouns receive an auxiliary "measure word" rather than a plural suffix. Unfortunately, English is not entirely consistent about countability and pluralization. DiYanni and Hoy (2001, their chapter 17) provide a good treatment.
- **Pronouns**. In some languages (including Chinese, Persian, Swahili, Turkish, and Finnish), the third-person pronouns are not gendered as they are in English: that is, there are no *he/she* or *his/her* pairs. (This is an enviable feature, actually; see http://scientistseessquirrel .wordpress.com/2015/04/24/dealing-with-the-defect-in-english.) Others (e.g., French, Italian, and Spanish) use gendered pronouns where English would use the nongendered *it* (for animals and objects). On a related note, some languages, such as Chinese, do not distinguish pronouns by case (no *she/her* or *I/me* distinction).
- **Word order**. English conventionally orders sentences subject/verb/ object, and places adjective before nouns, but other languages use different conventions. For example, Arabic and Hebrew use verb/ subject/object, while Hindi, Turkish, and Japanese use subject/object/verb; in French and Spanish, adjectives follow nouns.

- **Confrontational vs. deferential claims**. Speakers of Japanese (Gosden 1996), among others, may avoid confrontational ways of framing claims. EAL writers with this tendency may hesitate to point out knowledge gaps in their Introductions (see chapter 10), or may seem to downplay disagreement with previous studies in their Discussions (see chapter 13). Along similar lines, Sudanese writers avoid claims that rest only on their own data (ElMalik and Nesi 2008; it's not clear whether this tendency is shared by other speakers of Arabic). In contrast, some Spanish speakers use language that English speakers see as overclaiming (Englander 2009).
- **Writer-responsible and reader-responsible languages**. English is a "writer-responsible" language, in that the convention is for the writer to reduce the burden on the reader by providing substantial metadiscourse (writing about writing, as in "In this section we show" or "Next"), transitional expressions linking paragraphs and sections (see chapter 17), and the like. Languages such as Chinese, Japanese, Korean, Spanish, and German are more "reader-responsible," with readers expected to put in more effort to interpret the text. Speakers of these languages sometimes find conventional English metadiscourse "unsophisticated and childish" (Hanauer and Englander 2013).

Plagiarism, Culture, and Language

An extensive literature suggests that EAL writers are more likely to plagiarize than are native-English writers. While most of this literature focuses on writing in undergraduate coursework, there are at least anecdotes to the effect that the pattern extends to scientific writing for publication. This is an important issue, because penalties for plagiarism can be severe, including retraction of papers, expulsion from graduate schools, or even (occasionally) revocation of tenured appointments.

There may be several reasons that EAL writers have more difficulty avoiding plagiarism than native English speakers (Hayes and Introna 2005, Abasi and Graves 2008). First, lack of confidence in their writing skill may lead them to borrow phrasing from English-language papers (especially because the professional stakes are high, and the need to write in English exacerbates the demands of succeeding in a scientific career). Second, some EAL writers may be unfamiliar with Western publication practices, including the dividing line between paraphrasing and plagiarism. Third, for some EAL writers the issue is not just unfamiliarity but a genuine difference in practice: for example, Chinese writers may think of using another's words as a mark of respect rather than plagiarism (Pennycook 1996, Flowerdew 2007). To be clear, though, these are explanations for the apparently higher incidence of plagiarism by EAL writers, not reasons for maintaining the practice. As an EAL writer, you must take particular care to become familiar with norms for plagiarism in the scientific literature, and to apply extra vigilance so you don't inadvertently violate those norms. In particular, consulting with a more experienced English speaker, and, when appropriate, providing both modeled and new text, could help you identify writing that might be interpreted as plagiarism.

Writing Assistance

All writers gain from the help of others, but EAL writers in particular can reap enormous benefits when they get writing help from more fluent speakers of English. You might get such help from coauthors, friendly reviewers (chapter 22), writing consultants, or professional editors (among other options).

Reliance on native English coauthors is an obvious strategy, offering both immediate writing fixes and an opportunity for you to learn from their revisions. However, few EAL writers will have native English coauthors on every writing project. Friendly review (chapter 22) by a native English speaker (who is a scientist but not a coauthor) can be extremely helpful. If you're not yet proficient in English, though, keep in mind that it may be asking a lot to expect a friendly reviewer to provide substantial language revision. An alternative is to seek assistance from writing tutors or instructors available from campus writing centers (usually paid by the institution) or from professional editing services (paid by the writer). Some professional services are associated with major publishers (for instance, Elsevier and the Nature Publishing Group), while others are independent businesses. If you're considering writing center or fee-for-service help, ask whether the language expert is also familiar

with the scientific discipline. There is great value to this double expertise, which is more likely to be available in services associated with scientific publishers. Costs vary, but if professional editing reduces your
writing effort and makes publication more likely, this may be a good
investment.

Because EAL writers rank coauthors and friendly reviewers very
highly as sources of writing help (e.g., Uzuner 2008), anything that
makes coauthorship more likely or friendly reviewers more available
should pay dividends. This points to the importance of international
networking: if you can, attend English-language workshops and conferences (and while there, work to mingle with English-language scholars).

Submission and Review

EAL writers report a wide variety of experiences with journal submission and the review process (Hanauer and Englander 2013). Some journals actively work to increase international representation and therefore
may be especially welcoming to EAL writers. Others may be inadvertently unwelcoming—for example, by maintaining editorial boards entirely without EAL membership. This may be one consideration when
you choose a journal, although it shouldn't trump subject-matter match.

You may be tempted to see journal reviewers as a source of writing
help. This is unwise, because unlike friendly reviewers, writing center
experts, and paid editors, journal reviewers play a gatekeeping role in
addition to their manuscript-improvement role (chapter 23). Relying
too much on journal reviewers to improve the manuscript risks their
simply recommending rejection.

Many EAL scientists suspect bias in the review process against authors based on their names or their affiliations with non-English-
speaking institutions (Salager-Meyer 2008). It's certainly true that reviewers often point to English errors, suggest they arise because of an
author's EAL background, and ask for them to be fixed. By itself, this is
evidence only of a journal's desire to publish well-written papers, not of
reviewer bias. What's unsettling is the possibility that reviewers might
actually hold EAL writers to a higher standard than native-English ones.
Whether reviewers are actually biased this way is not clear, and some

may actually be more forgiving when they realize errors reflect the author's EAL status. Nonetheless, if you're worried about potential reviewer bias, you might prefer double-blind review (in which reviewers are not told the authors' names or affiliations) when it's available. Unfortunately, this will avoid only the most egregious biases—those operating even when the text is perfect—because removing names and affiliations won't prevent an experienced reviewer from guessing at your language background based on the kind of errors they see.

All this may sound as if EAL writers face huge obstacles in the review process. I don't think that's necessarily true. Most of the reviewers I've worked with as an editor, and most of the editors I know, understand the challenges EAL writers face, and want to help them publish their excellent work.

Chapter Summary

- The large majority of the scientific literature is now published in English, and English publications are more broadly indexed and cited.
- EAL writers should avoid writing in their native language for later translation.
- EAL writers can improve their English abilities in many ways, including reading, practice writing, working at English-language institutions, writing theses and presentations in English, using available English texts as models, and taking courses.
- EAL writers from different language backgrounds have different challenges, which arise from the particular differences between their native languages and English.
- The avoidance of plagiarism may be a particularly important issue for EAL writers, owing to lack of confidence, unfamiliarity with publishing practices, and cultural differences.
- Assistance for EAL writers can come from sources including cowriters, friendly reviewers, writing consultants, or professional editors. Journal reviewers should not be relied on for writing help.

Part VII

||||||||||||||||||||||||||||||||

Final Thoughts

Writing this book has been an interesting experience. (Also exciting, frustrating, energizing, agonizing, and all kinds of other adjectives.) At the beginning, I thought I knew certain things about scientific writing, and that writing the book would simply involve finding clear ways to express them. In light of my own arguments about the story evolving as it's told (chapters 5, 7, 21), I shouldn't have been surprised to find out that I was wrong. I learned a great deal about writing, and about myself as a writer, while working on this book.

Most of what I learned about writing fit well into the chapters I had already planned, but one thing did not. About halfway through the writing, I noticed a gaping hole in my thinking. I had been busily pushing my reader (you!) into a relentless drive to polish scientific writing for crystal clarity. But I hadn't thought much about possible side effects of that. In particular, if we value clarity so obsessively, is the product inevitably so utilitarian that it's easy to read, but colorless? Or can a reader experience pleasure in reading scientific writing that's clear but that also includes touches of whimsy, humor, or beauty? Coming to a tentative answer has involved a lot of reading and a lot of conversation with friends and colleagues. What resulted will serve as my final thoughts on writing. They are personal, they are likely to be controversial, and they may quite possibly be wrong. Nonetheless, I conclude with some thoughts on the potential for pleasure in scientific writing.

TWENTY-EIGHT

On Whimsy, Jokes, and Beauty:
Can Scientific Writing Be Enjoyed?

I've spent much of this book hammering at the notion that scientific writing should be crystal-clear, even "telepathically" clear (and suggesting ways that you can make it so). In chapter 1, I quoted approvingly from a number of writers and rhetoricians who endorse this idea, including Nathaniel Hawthorne, who asserted that the best style makes "the words absolutely disappear into the thought" (Van Doren 1949, 267). But wait: If the words disappear, leaving only the thought, they have done their job; but has an opportunity been squandered? Could your reader, in addition to effortlessly understanding your content, also enjoy the words that express it—perhaps because you've included small touches of whimsy, humor, or beauty? Could such enjoyment even enhance your communication with your reader?

When writing this book brought me unexpectedly to these questions, I was flummoxed. No book I was familiar with said anything about whether scientific writing can be made enjoyable, or about whether it should be[1]. I'd never been part of any conversation about this—not with a journal editor, not with a collaborator, not in a lab meeting critiquing published papers. I've since discovered that many of my colleagues have opinions on the subject. By and large, though, they keep these opinions to themselves (the exceptions being a few thoughts expressed in passing in Prefaces and Notes attached to publications about other things; e.g., Rosenzweig 1995, xv; Mangel 2006, xi).

---

[1] Since then, I've discovered some commentary in Harmon and Gross (2007), although it's more descriptive than prescriptive. Sword (2012) argues that academic writing can be "stylish," and includes some examples from scientific writing along with the humanities. Many books explore stylishness and beauty in expository writing more generally; Williams (1990) is well worth reading even though it doesn't deal with scientific writing in particular.

I've also learned that the question, unanswered, has been with us for hundreds of years. Think back to the roles of Thomas Sprat and Robert Boyle in developing the modern, clearly written scientific paper (chapter 1). As it happens, Sprat and Boyle disagreed somewhat about writing style. Sprat railed against anything more than the plainest language, with particular disdain for figures of speech such as metaphors, similes, or irony (collectively, "tropes"):

> Who can behold without indignation how many mists and uncertainties these specious Tropes and Figures have brought to our knowledge? How many rewards . . . have been . . . snatched away by the easy vanity of fine speaking? (Sprat 1667, 112; spelling and punctuation modernized)

This position would fit comfortably into most modern writing guides—including the preceding chapters of this one. But Boyle thought Sprat wanted to swing the pendulum too far:

> . . . to affect needless rhetorical ornaments in setting down an experiment . . . little less improper than . . . to paint the eyeglasses of a telescope . . . in which even the most delightful colors . . . would hinder the sight. . . . And yet I approve not that dull and insipid way of writing, which is practiced by many . . . for though a philosopher need not be solicitous that his style should delight his reader with his floridness, yet I think he may very well be allowed to take a care that it disgust not his reader by its flatness . . . Though it were foolish to color . . . the glasses of telescopes, yet to gild . . . the tubes of them may render them most acceptable to the users. (Boyle 1661, 11–12; spelling and punctuation modernized)

Could Boyle have had a point? There is no doubt that the first half of his argument is correct: in scientific writing, art (the attempt to make writing enjoyable) is a bad idea if it interferes with easy understanding (metaphorically, painting the telescope's eyeglass). In literature, beautifully enigmatic writing may be a success—as in the intentionally mysterious symbolism of Herman Melville's great white whale, Moby-Dick (Melville 1851). In science, beautifully enigmatic writing is a failure. This is why the previous chapters hammer so persistently at the notion of clarity: any thought of pleasure from writing must begin only once clarity has been achieved.

Boyle continues, though, with an argument that art, when it *doesn't* interfere with clarity (gilding the telescope's tube), can make scientific writing more effective by recruiting and retaining readers. Is this part of his argument right? I think so.

As an example, consider Hurlbert (1984): a long and difficult paper about pseudoreplication, a class of statistical problem that plagues ecology (among other fields). Pseudoreplication is an important issue for ecologists, but even statistics junkies like me don't feel frissons of excitement just thinking about it. In fact, twenty-five pages on pseudo-replication could easily be the statistical equivalent of watching very complicated paint dry. And yet Hurlbert's paper is still widely read and has been cited over four thousand times. I suspect that this success owes a lot to Hurlbert's discussion of the distinction between *demonic intrusion* and *nondemonic intrusion*[2] in experiments. I'll confess that the first time I read the paper, I stuck with it only because of this arresting terminology; and I recently came to reread it because the phrase "demonic intrusion" was nudging at my memory, demanding to be tracked down. Hurlbert's whimsy drew me through his important paper, drew me back to it, and has done the same for many other readers over the years.

While I claim that Hurlbert's impact has been enhanced by the reader's amusement with "demonic intrusion," I can't offer a formal analysis to prove this (or my more general claim that reader pleasure can enhance the function of scientific writing). Such an analysis would make an excellent Ph.D. thesis in science studies and I hope that someday, someone will conduct it. In the meantime, I can offer a look at some ways that scientific writing can tickle its reader's fancy. As we proceed, please ask yourself this question: would you rather read writing that's clear and functional but also fun to read, or writing that's clear and functional, but nothing more?

---

[2] What Hurlbert meant by *demonic intrusion* was a situation in which, for some reason, an unexpected but consistent perturbation is applied to every replicate of one experimental treatment but not the others—something a malevolent demon might do (although Hurlbert gives natural examples too). "Confounding" is the more customary term. *Nondemonic intrusion* is the occurrence of random perturbations affecting replicates of every treatment ("error" and "noise" are the customary terms).

Sightings (1): Playfulness in the Scientific Literature

I've called this section "*Sightings*" because the scientific literature has a reputation for being turgid and tedious—and while this may not be entirely deserved, it's not entirely *un*deserved either. Examples of sparkling or amusing scientific writing are unusual enough that finding one is like sighting a tapir or a narwhal: something to tell your friends about. Or perhaps tapirs and narwhals aren't the right metaphor, because while rare, these are big, conspicuous animals. Maybe a better one is a glow-throated hummingbird or a Salt Creek tiger beetle: beautiful, but tiny and glimpsed in passing. In scientific writing, pleasure for the reader tends to come from small things: playful touches, deft turns of phrase, allusions, and sparkling metaphors that ornament but don't dominate the text. When they're successful, such small touches don't compromise clarity; sometimes, they may even enhance it.

A famous example of whimsy (and literary allusion) is Gell-Mann's (1964) use of the name "quark" for a hypothesized new class of elementary particles. Gell-Mann (1964, 1995) attributed the name to a typically inscrutable line in James Joyce's *Finnegans Wake*: "Three quarks for Muster Mark!" (Joyce 1939, 383). Since three quarks make up each proton or neutron, the name fits. Interestingly, quarks were hypothesized independently by George Zweig (1964), in a paper published a couple of weeks before Gell-Mann's, but Zweig called them "aces." Quark, not ace, stuck, and Gell-Mann's prediction, not Zweig's, is widely remembered. There may be several reasons for this, but pleasure in Gell-Mann's naming was surely a factor.

Playful terminology can be found across disciplines. Systematists naming new genera and species have dubbed a suitably-shaped mite *Darthvaderum* after the *Star Wars* villain Darth Vader (Hunt 1996), a trap-door spider *Aname aragog* after Aragog, the spider from *Harry Potter and the Chamber of Secrets* (Harvey et al. 2012), and a sponge-shaped mushroom *Spongiforma squarepantsii* after the cartoon character SpongeBob SquarePants (Desjardin et al. 2011). Chemists play this game, too; for instance, a class of very large cyclic organic molecules has been dubbed the "dogcollaranes" (Craig and Paddon-Row 1987). But perhaps the best examples are in the naming of genes in *Drosophila*, the fruit fly model central to molecular and developmental genetics. With thousands of genes to name, and facing the alternative of having to remember names such as snoRNA:Me18S-G1358b, *Drosophila* geneticists have dubbed genes *saxophone, scarface, skittles, smaug, sneaky*, and *spotted dick* (to pull just a few examples from the S's).

Playful touches aren't limited to terminology. They are probably most common in titles and acknowledgements, albeit for different reasons. Playful titles are an attempt to catch the reader's attention; consider, for instance, Coleman and Lee's (1989) "Escape from the menace of the giant wormholes" or Rissman et al.'s (1999) "Sex with knockout models: behavioral studies of estrogen receptor alpha." (A "knockout model" is a mouse strain with a particular gene inactivated, or knocked out; and the estrogen receptor in question turned out to be involved in copulatory behavior.) Who wouldn't read on? Playfulness in acknowledgements is probably more of a relief valve for writers, who can write things there that wouldn't survive the editorial process elsewhere (as when Van Valen [1973] thanked a granting agency "for regularly rejecting my . . . grant applications for work on real organisms . . . thus forcing me into theoretical work").

Despite Sprat's aversion, scientific writers continue to use metaphors and similes. Some have become so familiar that we don't remember that when first used, they glistened ("tree of life," "big bang," "electron cloud"). Apt allusions to literature (and other arts) offer tinges of pleasure, too. Some become adopted as standard parts of discourse because they suit the science so well, as in evolutionary biology's Red Queen Hypothesis[3]. Others are simply decorations, as in the use of a quotation from T. S. Eliot's *Four Quartets* (Eliot 1943) to conclude the paper first reporting the sequencing of the human genome (International Human Genome Sequencing Consortium 2001):

---

[3] This is the notion that organisms must keep evolving just to maintain constant fitness as their environment changes and their enemies evolve better attacks. The term refers to the Red Queen in *Through the Looking-Glass*, the sequel to *Alice's Adventures in Wonderland*. The Red Queen and Alice run as fast as they can, and when Alice points out that they haven't left the spot they started in, the Red Queen explains "Now, *here*, you see, it takes all the running *you* can do, to keep in the same place" (Carroll 1871, 42).

We shall not cease from exploration
And the end of all our exploring
Will be to arrive where we started
And know the place for the first time.

This conveys nicely the satisfaction and the excitement of reporting a genome sequence that was a major accomplishment in its own right, but also a first step in a long journey toward understanding the biology underpinning our identity as humans.

The visual elements of scientific papers can also incorporate playfulness: decorative elements or visual jokes that occupy what might otherwise be white space. Writers who indulge in this practice are following a long tradition, dating back at least to medieval maps decorated with fanciful creatures like sea monsters and unicorns (Figure 28.1). Unicorns

Figure 28.1. A cavorting aquatic unicorn decorating an early chart of Alaska (Jode 1593). Rare Map G4371 S1 [1593?] D4, Alaska and Polar Regions Collections and Archives, Elmer E. Rasmuson Library, University of Alaska Fairbanks (cropped image).

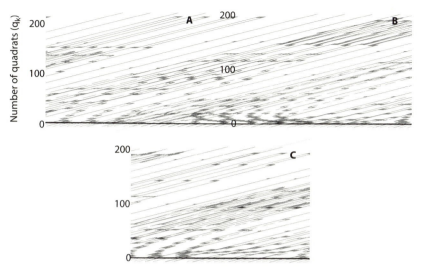

Figure 28.2. The habitat of the montane unicorn includes high mountains worldwide and also, apparently, bar charts. (modified from Hurlbert 1990).

still make the occasional appearance. Hurlbert (1990) wrote his paper "Spatial Distribution of the Montane Unicorn" to make a serious point about the detection of nonrandom spatial pattern in ecology. Since his (invented) unicorns lived in mountains, he let them climb the bar charts illustrating the shapes of their distributions (Figure 28.2). Wilson and Calvin (1955) used decoration more subtly. Their paper identified some intermediates in the chemical reactions of photosynthesis (as part of a larger research program for which Calvin won the 1961 Nobel Prize in Chemistry). The experiments used some elaborate apparatus to control flows of $^{14}CO_2$ and $^{12}CO_2$ across a vessel containing algal culture. They provided a schematic diagram of this apparatus, which upon close inspection reveals an unexpected feature of the algal vessel (Figure 28.3).

One could accuse Cornelius de Jode, who put the unicorn on the map, of painting the telescope's eyeglass, since many readers in his day would have taken his unicorn's possibility quite seriously. Hurlbert's unicorns, and Wilson and Calvin's fisherman, safely gild the telescope's tube, as there's no risk of readers confusing the ornaments with real fea-

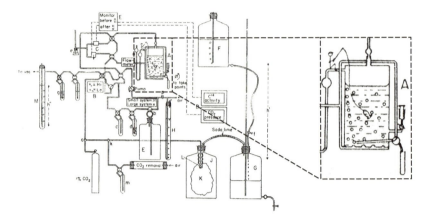

Figure 28.3. An apparatus for studying photosynthetic biochemistry. Modified from Wilson and Calvin (1955).

tures of data or apparatus. There's also no risk of readers escaping either paper without delight.

Sightings (2): Beauty

In reading the last section, perhaps you found yourself thinking that I was going after the easy pickings. Sure, scientists indulge in witty titles, playful terminology, and literary allusions; but what about truly beautiful prose? I will admit that *this* kind of sighting is rarer. This is in part because examples are rarer, although it doesn't help that nobody draws them to your attention by compiling them.

That one can write with grace and beauty *about* science is in no doubt. A long tradition of lay and popular science writing has included luminaries ranging from Henry David Thoreau to Rachel Carson, Carl Sagan, and John McPhee. I'll offer just one example: a passage from my favorite science essayist, Lewis Thomas, here writing about the evolutionary role of mutation:

> The capacity to blunder slightly is the real marvel of DNA. Without this special attribute, we would still be anaerobic bacteria and there would be no music . . . Each of the mutations that have brought us along represents a random, totally spontaneous accident, but it is no accident at all that mutations occur; the molecule of DNA was ordained from the beginning

to make small mistakes. If we had been doing it, we would have found some way to correct this, and evolution would have been stopped in its tracks . . . But there it is: we are here by the purest chance, and by mistake at that. Somewhere along the line, nucleotides were edged apart to let new ones in; maybe viruses moved in, carrying along bits of other, foreign genomes; radiation from the sun or from outer space caused tiny cracks in the molecule, and humanity was conceived (Thomas 1979, 23).

This passage is actually quite technical, and yet in addition to making it crystal-clear, Thomas makes it sing.

But writing about science for laypeople and writing science for the primary literature are two different things. Can one report new science with the kind of beauty Thomas and others manage in lay essays? Certainly it was possible in times gone by, when the literature was less crowded and we had the luxury of expansive writing in long papers and monographs. Here, famously, is Darwin concluding the *Origin of Species*:

It is interesting to contemplate a tangled bank, clothed with many plants of many kinds, with birds singing on the bushes, with various insects flitting about, and with worms crawling through the damp earth, and to reflect that these elaborately constructed forms, so different from each other, and dependent upon each other in so complex a manner, have all been produced by laws acting around us . . . There is grandeur in this view of life, with its several powers, having been originally breathed into a few forms or into one; and that, whilst this planet has gone cycling on according to the fixed law of gravity, from so simple a beginning endless forms most beautiful and most wonderful have been, and are being, evolved. (Darwin 1859, 489–90)

Can we pull off anything like this today, with the demands on journal space and on readers' attention that lie behind my call for clarity and brevity in writing? As I marshalled my thoughts to write this chapter, this certainly seemed the hardest question.

An interesting place to start toward an answer might be with Vladimir Nabokov. Nabokov, a literary giant, displayed "mastery of the aesthetic potential of prose" in writing *Lolita* and other novels that are among "the most carefully crafted works of verbal art in the twentieth century" (Morris 2010, 3). But he was also a scientist: in particular, a systematist who published papers on butterfly taxonomy. The bulk of

Nabokov's scientific writing consists of species descriptions and taxonomic revisions, which is a highly technical business depending on exhaustive detail of morphology and geographic distributions. So perhaps it's no shock that his butterfly papers aren't lyrical in every sentence from Abstract to Conclusion. His writing, instead, is sprinkled with small touches of art and humor, and reading them is an adventure that yields little sparkling surprises like these:

> [The genus *Cyclargus* has a sagum[4] with] two convex ... lobes ... connected at the zone with the alula, and below the zone with the points of the furca, converging in front ... of the adeagus in the manner of a stiffly bulging short waistcoat, too ample as it were for the body it encloses. (Nabokov 1952, 15)

and

> [The genus *Lycaeides*'] cradle is a lost country of plenty beyond the Arctic Circle ... ; its nurseries are the mountains of Central Asia, the Alps, and the Rockies. Seldom more than two and never more than three species are known to occur in a given geographical region, and so far as records go, not more than two species have ever been seen frequenting the same puddle or the same flowery bank. (Nabokov 1944, 111)

This pattern of clear and functional writing decorated with occasional glints of beauty is, I believe, the pattern we should be aiming for. Nabokov gilds the telescope's tube.

Perhaps you think it's easy to write lyrically about butterflies (even their genitalia). Maybe; so let's move on to quantum mechanics. Mermin (1995) discusses a startling result that seems to let an observer measure a set of properties that shouldn't be simultaneously measurable. Mermin calls the observer "Alice" and the measurement the "VAA trick" (after the authors who originally proposed it). In summing up the reason the VAA trick is possible, Mermin writes:

> There are no physical grounds for insisting that [Alice] assign the same value to an observable for each mutually commuting trio it belongs to—a

---

[4] *Sagum*, like *alula*, *furca*, and *adeagus*, refers to a part of the butterfly penis. Sorry. The male genitalia of insects are both complex and important in insect taxonomy, and Nabokov has to use technical terminology for these structures. His prose sparkles anyway.

requirement that would indeed trivially make her job impossible. The manner in which the nine-observable BKS theorem brings Alice to grief is more subtle than that. It is buried deep inside the mathematics that underlies the construction that makes it possible, when it is possible, to do the VAA trick. (Mermin 1995, 834)

You don't have to understand this passage to appreciate the lovely writing.

Rockman (2012) made particularly effective use of an extended and striking metaphor in a paper reviewing the search for mutations that have large effects on the traits of organisms (so-called "quantitative trait nucleotides," or QTNs). This is an important problem in quantitative genetics, because mutations (alleles) of large effect are easy to detect, but may be uncommon; most evolutionary change may instead be due to the accumulation of many alleles each making a tiny contribution to change. In introducing the problem, Rockman writes:

> In January 1848, James Marshall found gold flakes in the millrace of John Sutter's saw mill. Within months ... the rush was on. Thousands left home, rounding the Cape, crossing the Isthmus, or joining the wagon trains headed west. Soon the easy pickings were gone, and consortia of miners banded together to blast more flakes from the hills. Extraction technologies proliferated: first rockers and long toms, then gravel dredges, and finally hydraulic mining, which washed whole mountains through giant sluices to recover dense gold flakes from the riffles.
>
> Modern day QTN prospecting is the Sierra Nevada of the 1850s. The shiny (Mendelian) nuggets are rapidly being collected, and ever larger teams of researchers with ever more powerful technologies are now probing whole genomes to find their quarry. But visible flakes of placer gold represent a small fraction of the global gold reserve; most gold is in microscopic particles concealed in low-grade ore. If the stuff of evolution is often alleles of microscopic effect, large-effect nuggets can tell us little about the material basis for evolution. (Rockman 2012, 2)

No frequency histogram could make Rockman's point as vividly as the gold metaphor: large-effect alleles are easy to find, and exciting when you find one, but they might not contribute much to overall genetic variation.

The last word can go to chemistry. Kroto et al. (1985) name, and describe the likely structure of, a novel molecule in a paper titled simply "C_{60}: Buckminsterfullerene"[5]. Here they describe their proposed structure, a soccer ball of carbon atoms:

> An unusually beautiful (and probably unique) choice is the truncated icosahedron... All valences are satisfied with this structure, and the molecule appears to be aromatic. The structure has the symmetry of the icosahedral group. The inner and outer surfaces are covered with a sea of π electrons (Kroto et al. 1985, 162).

Two things unite these examples. First, they achieve touches of beauty without compromising clarity. Second, they let us share the joy the writers must have felt in drawing back Nature's curtains to make their new discoveries. Most of us feel this joy from time to time; why not let it show in our writing?

Pushback

Perhaps you don't share my admiration for amusing, playful, or beautiful scientific writing. If so, you are not alone: there is plenty of pushback against the idea of pleasure in our literature. The *Journal of Organic Chemistry* once published a paper written almost entirely in blank verse (Bunnett and Kearley 1971), but in a footnote, the editors made it quite clear that they didn't envision doing so again. Seringhaus et al. (2008) and others have criticized "silly" gene names. Gross et al. (2002, 167) give some examples of writing showing "personal style," and opine that these are "examples of 'bad' scientific prose... if they were far more pervasive they would be counterproductive in communicating science effectively."

I ran into this pushback myself in submitting a manuscript about the population genetics of an endangered plant (Heard et al. 2009). The plant seems to self-pollinate despite having many florets lacking stamens (the male structure), and we thought this likely happened by me-

---

[5] The name honors the architect and futurist Buckminster Fuller, who used similar geometries to design graceful buildings. Few chemical names roll of the tongue quite like "buckminsterfullerene."

chanical transfer of pollen from floret to floret as plants move in the wind. So I wrote "there is, however, considerable evidence (Houle 1988) for pollen transfer among florets . . . by wind or shaking (Hall et al. 1957)." An attentive reader would discover that the Hall et al. reference was to Jerry Lee Lewis's classic song "Whole Lotta Shakin' Goin' On." One reviewer noticed this, and groused, "Although I appreciated the levity from the reference Hall et al. (1957) I think it is not appropriate for a scientific publication." We had to drop the joke.

You can argue that my intended joke wasn't very funny (and I'd be hard pressed to defend myself). But that wasn't the reviewer's objection. Instead, the objection was that it wasn't "appropriate" for the scientific literature. What might that mean, and what lies behind the more general pushback I've described? I can see two possibilities. Some objections are Boyle's: the belief that a particular joke, metaphor, or bit of beautiful writing is obscuring meaning (painting the eyeglass). I surveyed the editorial board of a major journal in ecology and evolution about attitudes to pleasure in writing, and this was the most frequent refrain in their replies: worry that some readers wouldn't get a joke or understand a metaphor, and that this would impair communication. This objection to *particular* attempts at pleasure in scientific writing could presumably be overcome if the metaphor was very accessible or the joke just wouldn't be noticed by those who didn't get it. Other objections seem to be that scientific writing isn't supposed to be funny or beautiful *in general*. As Sagi and Yechiam (2008) put it, perhaps it's that "traditionally, scientific publication is considered a serious matter, and humor seems antithetical to it" (p. 686). I think that was the position of the reviewer who nixed my Jerry Lee Lewis joke, and there isn't much doubt that many scientists think this way—and that you'll meet them as reviewers and as editors. Personally, I think this is too bad.

A Prescription

If all you want to do is write with crystal clarity, with prose that achieves its primary function of being effortless to read, that's great: doing so will place your work among the very best of the scientific literature. But if you want to reach for even more, if you agree with me that we can also offer our readers some pleasure, what can you do? You can't suddenly decide to write all your papers in the form of stand-up comedy scripts or metaphor-rich sonnets—and you shouldn't. But you can try to write with small touches of whimsy, humor, and beauty—without, of course, compromising clarity; and even knowing that sometimes, reviewers will make you take them out. When you review manuscripts, you can suppress any reflex that tells you to question such touches; you can even (gently) suggest they be put in. Finally, you can announce your admiration of writing that has given you pleasure. Announce it to the writers who crafted the passage, to editors who might be considering its fate, and to students or colleagues who might read it. We can change our culture to deliver, and value, pleasure in our writing—if we choose to.

Of course, I have to close by circling back to the crucial importance of clarity. Your readers will appreciate your touches of whimsy, humor, and beauty only if they don't interfere with their easy understanding of what you're writing about. Clarity should always be the first thing on your mind as you write—and the last thing, too.

Chapter Summary

- The pursuit of clarity need not preclude touches of whimsy, humor and beauty in scientific writing.
- Humor and beauty may even enhance writing, by recruiting readers to a paper, retaining them as they read, and fixing it in their memory.
- Notwithstanding the appeal of whimsy, jokes, and beauty, clarity remains the most important attribute of good scientific writing.

||||||||||||||||||||||||||||| Acknowledgements |||||||||||||||||||||||||||||||||

For comments on this book's manuscript I'm grateful to Peter Abrams, Fred Allendorf, John Paul Balmonte, Madeline Bassnet, Alex Bond, Drew Carleton, Emilie Champagne, Stylianos Chatzimanolis, Leah Flaherty, Mischa Giasson, Dorthea Grégoire, Kristie Heard, Steve Hendrix, Mikaela Huntzinger, Viqar Husain, Linley Jesson, Rob Johns, Richard Kemick, Dave Magee, Randall Martin, Arne Mooers, Yolanda Moreby, Kim Nicholas, Tobi Oke, Peter Petraitis, Vlad Tasić, Shaopeng Wang, Bo Zhang, and three anonymous reviewers. Lonnie Aarsen, Fred Adler, Justin Ancheta, Judie Bronstein, Chris Darling, Jeremy Fox, Brock Harpur, Lina Kabbadj, Drew Kerkhoff, Marc Mangel, Trish Morse, Sean O'Donnell, Jay Stachowicz, and Peter Taylor offered comments (but not necessarily agreement) on the issue of humor and beauty in scientific writing. Chandra Moffat kindly allowed me to quote her pre- and post-revision Abstract in chapter 20, and Sanjeev Seahra explained some quantum mechanics to me. Josh Dickison, Fran Holyoke, Judy MacLean, Phil Taber, and others behind the scenes at the UNB Libraries provided outstanding library support; Kim Stafford and Zach Selley helped trace William Stafford's thoughts on writer's block. The University of New Brunswick's Department of English and the University of Toronto's Koffler Biological Reserve kindly provided space for me during the writing of this book. Finally, Jamie Heard and the makers of the *NOVA* documentary series inspired my ALMA Observatory/massive-star formation example.

My most important acknowledgements, though, are for all those scientists and educators who have helped me learn my own craft of writing—from grade school teachers all the way through the colleagues, reviewers, and friends who have read my most recent manuscripts. Any list I could attempt to offer would be enormously long but inevitably incomplete. If you belong on that list, thank you.

References

Aad G et al. (ATLAS Collaboration, CMS Collaboration) (2015) Combined measurement of the Higgs boson mass in pp collisions at $\sqrt{s}$ = 7 and 8 TeV with the ATLAS and CMS experiments. Physical Review Letters 114:191803

Aaij R et al. (LHCb collaboration) (2014) Updated measurements of exclusive J/ψ and $\psi(2S)$ production cross-sections in pp collisions at $\sqrt{s}$ = 7 TeV. Journal of Physics G: Nuclear and Particle Physics 41:055002

Abasi AR, Graves B (2008) Academic literacy and plagiarism: Conversations with international graduate students and disciplinary professors. Journal of English for Academic Purposes 7:221–233

Ackerman D (1990) A natural history of the senses. Random House, New York, NY

Akerlof G (1970) The market for lemons. Quarterly Journal of Economics 84:488–500

Alquist J, Baumeister RF (2012) Self-control: Limited resources and extensive benefits. WIREs Cognitive Science 3:419–423

American Institute of Physics Publication Board (1990) AIP style manual. 4th edition. American Institute of Physics, New York, NY

Ammon U (2012) Linguistic inequality and its effects on participation in scientific discourse and on global knowledge accumulation: With a closer look at the problems of the second-rank language communities. Applied Linguistics Review 3:333–355

Ancheta J, Heard SB (2011) Impacts of insect herbivores on rare plant populations. Biological Conservation 144:2395–2402

Ancheta J, Heard SB, Lyons JW (2010) Impacts of salinity and simulated herbivory on survival and reproduction of the threatened Gulf of St. Lawrence aster, *Symphyotrichum laurentianum*. Botany 88:737–744

Anholt RRH (1994) Dazzle 'em with style: The art of oral scientific presentation. W.H. Freeman, New York, NY

Ariga A, Lleras A (2011) Brief and rare mental "breaks" keep you focused: Deactivation and reactivation of task goals preempt vigilance decrements. Cognition 118:439–443

Arvey RD, Rotundo M, Johnson W, Zhang Z, McGue M (2006) The determinants of leadership role occupancy: Genetic and personality factors. Leadership Quarterly 17:1–20

Bacon F (1609) De sapientia veterum. In: Spedding J, Ellis RL, Heath DD (eds.) (1860) The works of Francis Bacon vol 8. Brown and Taggard, Boston, pp. 155–156.

Bacon F (1627) New Atlantis. Dr. Rawley, London, UK

Ballard JG (2003) Millennium people. Harper Collins, New York, NY

Bandura A (1997) Self-efficacy: The exercise of control. Freeman, New York, NY

Banerjee R, Pudritz RE (2007) Massive star formation via high accretion rates and early disk-driven outflows. Astrophysical Journal 660:479–488

Basturkmen H (2012) A genre-based investigation of discussion sections of research articles in dentistry and disciplinary variation. Journal of English for Academic Purposes 11:134–144

Beaver DD, Rosen R (1978) Studies in scientific collaboration. I. The professional origins of scientific co-authorship. Scientometrics 1:6–84

Bizarro JP (2013) Comment on "Wigner function for a particle in an infinite lattice." New Journal of Physics 15: 068001

Blum SD (2009) My word! Plagiarism and college culture. Cornell University Press, Ithaca, NY

Boice R (1990) Professors as writers: A self-help guide to productive writing. New Forums Press, Stillwater, OK

Boice R (2000) Advice for new faculty members: *Nihil nimus*. Allyn and Bacon, Boston, MA

Bonnell IA, Bate MR (2006) Star formation through gravitational collapse and competitive accretion. Monthly Notices of the Royal Astronomical Society 370:488–494

Borenstein M, Hedges LV, Higgins JPT, Rothstein HR (2009) Introduction to meta-analysis. John Wiley and Sons, Chichester, UK

Bowman LL, Levine LE, Waite BM, Gendron M (2010) Can students really multitask? An experimental study of instant messaging while reading. Computers & Education 54:927–931

Boyle R (1660) New experiments physico-mechanical: Touching the spring of the air and its effects, made, for the most part, in a new pneumatical engine. H. Hall, Oxford, UK.

Boyle R (1661) Certain physiological essays written at distant times, and on several occasions. Henry Herringman, London, UK

Boyle R (1665a) An account of a very odd monstrous calf. Philosophical Transactions of the Royal Society (London) 1:10

Boyle R (1665b) Observables upon a monstrous head. Philosophical Transactions of the Royal Society (London) 1:85–86

Bryan G, Karlan D, Nelson S (2010) Commitment devices. Annual Review of Economics 2:671–698

Bryson B (2004) Bryson's dictionary of troublesome words. Broadway Books, New York, NY

Bunnett JF, Kearley FJ (1971) Comparative mobility of halogens in reactions of dihalobenzenes with potassium amide in ammonia. Journal of Organic Chemistry 36:184–186

Calcagno V, Demoinet E, Gollner K, Guidi L, Ruths D, de Mazencourt C (2012) Flows of research manuscripts among scientific journals reveal hidden submission patterns. Science 338:1065–1069

Carrier J, Monk TH (2000) Circadian rhythms of performance: New trends. Chronobiology International 17:719–732

Carroll L (1871) Through the looking-glass, and what Alice found there. Macmillan, London, UK

Carter GG, Wilkinson GS (2013) Food sharing in vampire bats: Reciprocal help predicts donations more than relatedness or harassment. Proceedings of the Royal Society B: Biological Sciences 280: 20122573

Casadevall A, Fang FC (2010) Reproducible science. Infection and Immunity 78:4972–4975

Chabrier G (2003) Galactic stellar and substellar initial mass function. Publications of the Astronomical Society of the Pacific 115:763–795

Chaffer CL, Marjanovic ND, Lee T, Bell G, Kleer CG, Reinhardt F, D'Alessio AC, Young RA, Weinberg RA (2013) Poised chromatin at the ZEB1 promoter enables breast cancer cell plasticity and enhances tumorigenicity. Cell 154:61–74

Cho DW (2009) Science paper writing in an EFL context: The case of Korea. English for Specific Purposes 28:230–239

Chua HF, Ho SS, Jasinska AJ, Polk TA, Welsh RC, Liberzon I, Strecher VJ (2011) Self-related neural response to tailored smoking-cessation messages predicts quitting. Nature Neuroscience 14:426–427

Cicero (55 BC) De oratore. (1942) Harvard University Press, Cambridge, MA

CMS Collaboration, LHCb Collaboration (2015). Observation of the rare $B_s^0 \to \mu^+\mu^-$ decay from the combined analysis of CMS and LHCb data. Nature 522:68–72

Coleman S, Lee K (1989) Escape from the menace of the giant wormholes. Physics Letters B 221:242–249

Council of Science Editors SMC (2006) Scientific style and format: The CSE manual for authors, editors, and publishers. 7th edition. Council of Science Editors, Reston, VA

Craig DC, Paddonrow MN (1987) Crystal structures of 3 long, rigid, norbornylogous compounds of relevance to distance-dependence studies of long-range intramolecular electron-transfer processes. Australian Journal of Chemistry 40:1951–1964

Darwin C (1859) The Origin of Species. John Murray, London, UK

Daston L, Gallison P (2007) Objectivity. Zone Books, New York, NY

Davenas E, Beauvais F, Amara J, Oberbaum M, Robinzon B, Miadonna A, Tedeschi A, Pomeranz B, Fortner P, Belon P, Saintelaudy J, Poitevin B, Benveniste J (1988) Human basophil degranulation triggered by very dilute antiserum against IgE. Nature 333:816–818

Davis M (2005) Scientific papers and presentations. Revised edition. Academic Press, Burlington, MA

Day RA, Gastel B (2006) How to write and publish a scientific paper. 6th edition. Greenwood Press, Westport, CT

Derham W (1733) Letter to John Conduitt, 18 July 1733; Keynes Ms. 133, King's College, Cambridge, UK

Desjardin DE, Peay KG, Bruns TD (2011) *Spongiforma squarepantsii*, a new species of gasteroid bolete from Borneo. Mycologia 103:1119–1123

DiYanni R, Hoy PC II (2001) The Scribner handbook for writers. 3rd edition. Allyn and Bacon, Boston, MA

Eisenberger R (1992) Learned industriousness. Psychological Review 99:248–267

Eliot TS (1943) Four quartets. Harcourt Brace, New York, NY

ElMalik AT, Nesi H (2008) Publishing research in a second language: The case of Sudanese contributors to international medical journals. Journal of English for Academic Purposes 7:87–96

Englander K (2009) Transformation of the identities of nonnative English-speaking scientists as a consequence of the social construction of revision. Journal of Language, Identity, and Education 8:35–53

Fanelli D (2009) How many scientists fabricate and falsify research? A systematic review and meta-analysis. PLOS One 4:e5738

Feder ME, Mitchell-Olds T (2003) Evolutionary and ecological functional genomics. Nature Reviews Genetics 4:649–655

Finkl CW (series editor) (1968–2016) Encyclopedia of earth sciences. Springer, New York, NY

Fleischmann M, Pons S (1989) Electrochemically induced nuclear fusion of deuterium. Journal of Electroanalytical Chemistry 261:301–308

Flowerdew J (1999) Writing for scholarly publication in English: The case of Hong Kong. Journal of Second Language Writing 8:123–145

Flowerdew J (2007) The non-Anglophone scholar on the periphery of scholarly publication. AILA Review 20:14–27

Foote S (1958) The Civil War: A narrative. 1. Fort Sumter to Perryville. Random House, New York, NY

Foote S (1963) The Civil War: A narrative. 2. Fredericksburg to Meridian. Random House, New York, NY

Foote S (1974) The Civil War: A narrative. 3. Red River to Appomattox. Random House, New York, NY

Foote S (1994) Shelby Foote: Stars in their courses: The Gettysburg campaign. C-SPAN Booknotes, http://www.booknotes.org/FullPage.aspx?SID=60099–1

Fowler HR, Aaron JE (2011) The Little, Brown handbook. Pearson, Boston, MA

Fowler HW, Burchfield RW (1996) The new Fowler's modern English usage. Clarendon Press, Oxford, UK

Friedland A, Folt C (2009) Writing successful science proposals. 2nd edition. Yale University Press, New Haven, CT

Galak J, Nelson LD (2011) The virtues of opaque prose: How lay beliefs about fluency influence perceptions of quality. Journal of Experimental Social Psychology 47:250–253

Gell-Mann M (1964) A schematic model of baryons and mesons. Physics Letters 8:214–215

Gell-Mann M (1995) The quark and the jaguar: Adventures in the simple and the complex. Henry Holt, New York, NY

Glänzel W, Schubert A (2005) Analyzing scientific networks through co-authorship. In: Moed HF, Glänzel W, Schmoch U (eds) Handbook of quantitative science and technology research: The use of publication and patent statistics in studies of S&T systems. Kluwer, New York, NY, pp. 257–276

Godden DR, Baddeley AD (1975) Context-dependent memory in two natural environments: On land and underwater. British Journal of Psychology 66:325–331

Gopen GD, Swan JA (1990) The science of scientific writing. American Scientist 78:550–558

Gordin, MD (2015) Scientific Babel: How science was done before and after global English. University of Chicago Press, Chicago, IL

Gosden, H (1996) Verbal reports of Japanese novices' research writing practices in English. Journal of Second Language Writing 5:109–128

Gross AG, Harmon JE, Reidy M (2002) Communicating science: The scientific article from the 17th century to the present. Oxford University Press, Oxford, UK

Gulick S, Reece R, Christeson G, van Avendonk H, Worthington L, Pavlis T (2013) Seismic images of the Transition fault and the unstable Yakutat-Pacific-North American triple junction. Geology 41:571–574

Gustavsson JP, Weinryb RM, Göransson S, Pedersen NL, Åsberg M (1997) Stability and predictive ability of personality traits across 9 years. Personality and Individual Differences 22:783–791

Halverson KL, Heard SB, Nason JD, Stireman JO III (2008a) Differential attack on diploid, tetraploid, and hexaploid *Solidago altissima* L. by five insect gallmakers. Oecologia 154:755–761

Halverson KL, Heard SB, Nason JD, Stireman JO III (2008b) Origins, distribution and local co-occurrence of polyploids in *Solidago altissima* L. American Journal of Botany 95:50–58

Hanauer DI, Englander K (2013) Scientific writing in a second language. Parlor Press, Anderson, SC

Harmon JE, Gross AG (2007) The scientific literature: A guided tour. University of Chicago Press, Chicago, IL

Harvey FSB, Framenau VW, Wojcieszek JM, Rix MG, Harvey MS (2012) Molecular and morphological characterisation of new species in the trapdoor spider genus *Aname* (Araneae: Mygalomorphae: Nemesiidae) from the Pilbara bioregion of Western Australia. Zootaxa:15–38

Hayes N, Introna LD (2005) Cultural values, plagiarism, and fairness: When plagiarism gets in the way of learning. Ethics and Behavior 15:213–231

Heard SB (1992) Patterns in tree balance among cladistic, phenetic, and randomly generated phylogenetic trees. Evolution 46:1818–1826

Heard SB (1995) Short-term dynamics of processing chain systems. Ecological Modelling 80:57–68

Heard SB (2012) Use of host-plant trait space by phytophagous insects during host-associated differentiation: The gape-and-pinch model. International Journal of Ecology 2012:ID192345

Heard SB, Cox GH (2007) The shapes of phylogenetic trees of clades, faunas, and local assemblages: Exploring spatial pattern in differential diversification. American Naturalist 169:E107–E118

Heard SB, Jesson LK, Tulk K (2009) Population genetic structure of the Gulf of St. Lawrence aster, *Symphyotrichum laurentianum* (Asteraceae), a threatened coastal endemic. Botany 87:1089–1095

Heard SB, Kitts EK (2012) Impact of *Gnorimoschema* gallmakers on their ancestral and novel *Solidago* hosts. Evolutionary Ecology 26:879–892

Heard SB, Remer LC (1997) Clutch size behavior and coexistence in ephemeral-patch competition models. American Naturalist 150:744–770

Heard SB, Remer LC (2008) Travel costs, oviposition behavior and the dynamics of insect-plant systems. Theoretical Ecology 1:179–188

Hendrickson R (1994) The literary life and other curiosities. Revised edition. Houghton Mifflin, New York, NY

Higgs P (1964) Broken symmetries and the masses of gauge bosons. Physical Review Letters 13:508–509

Higham NJ (1998) Handbook of writing for the mathematical sciences. Society for Industrial and Applied Mathematics, Philadelphia, PA

Hind KR, Saunders GW (2013) A molecular phylogenetic study of the tribe Corallineae (Corallinales, Rhodophyta) with an assessment of genus-level taxonomic features and descriptions of novel genera. Journal of Phycology 49:103–114

Hobbes T (1655) Elementorum philosophiae sectio prima De corpore. The English works of Thomas Hobbes of Malmesbury (1839 edition). John Bohn, London

Hughes M (2011) Online comment, *The Globe and Mail* (Toronto); confirmed by pers. comm.

Hunt GS (1996) Description of predominantly arboreal platermaeoid mites from eastern Australia (Acarina: Cryptostigmata: Plateremaeoidea). Records of the Australian Museum 48:303–324

Hurlbert SH (1984) Pseudoreplication and the design of ecological experiments. Ecological Monographs 54:187–211

Hurlbert SH (1990) Spatial distribution of the montane unicorn. Oikos 58:257–271

Hyland K (1998) Hedging in scientific research articles. John Benjamins, Amsterdam

Idzik KR, Nodler K, Licha T (2014) Efficient synthesis of readily water-soluble amides containing sulfonic groups. Synthetic Communications 44:133–140

International Human Genome Sequencing Consortium (Lander ES et al.) (2001) Initial sequencing and analysis of the human genome. Nature 409:860–921

Janssen N (2013) Response exclusion in word-word tasks: A comment on Roelofs, Piai and Schriefers. Language and Cognitive Processes 28:672–678

Jode C de (1593) Speculum orbis terrae. Arnold Koninx, Antwerp

Johnson ED (1991) The handbook of good English. Washington Square Books, New York, NY

Joyce J (1939) Finnegans wake. Faber and Faber, London

Katz MJ (2006) From research to manuscript: A guide to scientific writing. Springer, Dordrecht

Khatib M, Moradian MR (2011) Deductive, inductive, and quasi-inductive writing styles in Persian and English: Evidence from media discourse. Studies in Language and Literature 2:81–87

King C (2013) Single-author papers: A waning share of output, but still provide the tools for progress. http://sciencewatch.com/articles/single-author-papers-waning-share-output-still-providing-tools-progress

King S (2000) On writing: A memoir of the craft. Scribner, New York

Kolaczan CR, Heard SB, Segraves KA, Althoff DM, Nason JD (2009) Spatial and genetic structure of host-associated differentiation in the parasitoid *Copidosoma gelechiae*. Journal of Evolutionary Biology 22:1275–1283

Kosslyn SM (2006) Graph design for the eye and mind. Oxford University Press, Oxford, UK

Kroto HW, Heath JR, O'Brien SC, Curl RF, Smalley RE (1985) C_{60}: Buckminsterfullerene. Nature 318:162–163

Kupfer JA, Webbeking AL, Franklin SB (2004) Forest fragmentation affects early successional patterns on shifting cultivation fields near Indian Church, Belize. Ecosystems and Environment 103:509–518

Li K, Qian S-B (2013) Two particular EA-type binaries in the globular cluster ω Centauri. Research in Astronomy and Astrophysics 13:827–834

Loscalzo J (2012) Irreproducible experimental results: Causes, (mis)interpretations, and consequences. Circulation 125:1211–1214

Lozano GA, Lariviere V, Gingras Y (2012) The weakening relationship between the impact factor and papers' citations in the digital age. Journal of the American Society for Information Science and Technology 63:2140–2145

Lyons J (2009) The house of wisdom: How the Arabs transformed western civilization. Bloomsbury, New York

Maddox GD, Cook RE, Wimberger PH, Gardescu S (1989) Clone structure in four *Solidago altissima* (Asteraceae) populations: Rhizome connections within genotypes. American Journal of Botany 76:218–326

Magnusson WE (1996) How to write backwards. Bulletin of the Ecological Society of America 77:88

Mangel M (2006) The theoretical biologist's toolbox. Cambridge University Press, Cambridge, UK

Martin Ginis KA, Bray SR (2010) Application of the limited strength model of self-regulation to understanding exercise effort, planning and adherence. Psychology and Health 25:1147–1160

Maupertuis P-LM de (1737) La figure de la terre, determinée par les messieurs de l'Académie Royale des Sciences, qui on mesuré le degré du méridien au Circle Polaire. Mémoires de l'Académie Royale des Sciences 1737:389–466

Mazur JE (1996) Procrastination by pigeons: Preference for larger, more delayed work requirements. Journal of the Experimental Analysis of Behavior 65:159–171

Meer L van der, Costafreda S, Aleman A, David AS (2010) Self-reflection and the brain: A theoretical review and meta-analysis of neuroimaging studies with implications for schizophrenia. Neuroscience and Biobehavioral Reviews 34:935–946

Melville H (1851) Moby-Dick; or, the whale. Harper and Brothers, New York, NY

Mermin ND (1995) Limits to quantum mechanics as a source of magic tricks: Retrodiction and the Bell-Kochen-Specker theorem. Physical Review Letters 74:831–834

Michel J-B, Shen YK, Aiden AP, Veres A, Gray MK, The Google Books Team, Pickett JP, Hoiberg D, Clancy D, Norvig P, Orwant J, Pinker S, Nowak MA, Aiden EL (2011) Quantitative analysis of culture using millions of digitized books. Science 331:176–182

Mok WE (1993) Contrastive rhetoric and the Japanese writer of EFL. JALT Journal 15:151–161

Montgomery SL (2003) The Chicago guide to communicating science. University of Chicago Press, Chicago, IL

Morris PD (2010) Vladimir Nabokov: Poetry and the lyric voice. University of Toronto Press, Toronto

Murray DM (1990) Shoptalk: Learning to write with writers. Boynton/Cook, Portsmouth, NH

Nabokov V (1944) Notes on the morphology of the genus *Lycaeides* (Lycaenidae, Lepidoptera). Psyche 51:104–138

Nabokov V (1952) Notes on neotropical Plebejinae (Lycaenidae, Lepidoptera). Psyche 52:1–61

Nason JD, Heard SB, Williams FR (2002) Host associated genetic differentiation in the goldenrod elliptical-gall moth, *Gnorimoschema gallaesolidaginis* (Lepidoptera: Gelechiidae). Evolution 56:1475–1488

Nicholas KA, Gordon W (2011) A quick guide to writing a solid peer review. Eos 92:233–240

Novak JD, Cañas AJ (2008) The theory underlying concept maps and how to construct and use them. Technical Report IHMC CmapTools 2006–01 Rev 01-2008. Florida Institute for Human and Machine Cognition, Pensacola, FL.

Oke OA, Heard SB, Lundholm JT (2014) Integrating phylogenetic community structure with species distribution models: An example with plants of rock barrens. Ecography 37:614:625

Oppenheimer DM (2006) Consequences of erudite vernacular utilized irrespective of necessity: Problems with using long words needlessly. Applied Cognitive Psychology 20:139–156

Oruç A (2012) A handbook of scientific proposal writing. CRC Press, Boca Raton, FL

Osycka-Salut C, Diez F, Burdet J, Gervasi MG, Franchi A, Bianciotti LG, Davio C, Perez-Martinez S (2014) Cyclic AMP efflux, via MRPs and A1 adenosine receptors, is critical for bovine sperm capacitation. Molecular Human Reproduction 20:89–99

Palmer D, Digges E (1665) An extract of a letter containing some observations, made in the ordering of silk-worms, communicated by that known vertuoso, Mr. Dudley Palmer, from the ingenuous Mr. Edward Digges. Philosophical Transactions of the Royal Society (London) 1:26–27

Peacock M (2002) Communicative moves in the discussion section of research articles. System 30:479–497

Pecorari D (2008) Academic writing and plagiarism: A linguistic analysis. Continuum, London, UK

Pennycook A (1996) Borrowing others' words: Text, ownership, memory, and plagiarism. TESOL Quarterly 30:201–230

Pérez-Ramos A (1996) Bacon's legacy. In: Peltonnen M (ed) The Cambridge companion to Bacon. Cambridge University Press, Cambridge, UK, pp. 311–334

Pinker S (2014) The sense of style: The thinking person's guide to writing in the 21st century. Viking, New York

Queenan J (2009) Newman, Hoffman, and me. *The Guardian* (London), Apr 25

Quiller-Couch A (1916) On the art of writing: Lectures delivered in the University of Cambridge. Cambridge University Press, Cambridge, UK

Ramsden JJ (2009) Impact factors: A critique. Journal of Biological Physics and Chemistry 9:139–140

Rissman EF, Wersinger SR, Fugger HN, Foster TC (1999) Sex with knockout models: Behavioral studies of estrogen receptor alpha. Brain Research 835:80–90

Robinson JW, Skelly Frame EM, Frame GM II (2005) Undergraduate instrumental analysis. 6th edition. Marcel Dekker, New York, NY

Rockman MV (2012) The QTN program and the alleles that matter for evolution: All that's gold does not glitter. Evolution 66:1–17.

Rosenzweig ML (1995) Species diversity in space and time. Cambridge University Press, Cambridge, UK

Rubinstein A (2003) "Economics and psychology"? The case of hyperbolic discounting. International Economic Review 44:1207–1216

Sagi I, Yechiam E (2008) Amusing titles in scientific journals and article citation. Journal of Information Science 34:680–687

Salager-Meyer F (2008) Scientific publishing in developing countries: Challenges for the future. Journal of English for Academic Purposes 7:121–132

Sanders DB, Scoville NZ, Solomon PM (1985) Giant molecular clouds in the galaxy. 2. Characteristics of discrete features. Astrophysical Journal 289:373–387

Saunders GW, Clayden SL (2010) Providing a valid epithet for the species widely known as *Halosacciocolax kjellmand* S. Lund (Palmariales, Rhodophyta)—*Rhodophysema kjellmanii* sp. nov. Phycologia 49:628

Seringhaus MR, Cayting PD, Gerstein MB (2008) Uncovering trends in gene naming. Genome Biology 9:401–404

Shapin S (1984) Pump and circumstance: Robert Boyle's literary technology. Social Studies of Science 14:481–520

Silvia P (2007) How to write a lot: A practical guide to productive academic writing. American Psychological Association, Washington, DC

Soutullo A, Dodsworth S, Heard SB, Mooers AO (2005) Distribution and correlates of carnivore phylogenetic diversity across the Americas. Animal Conservation 8:249–258

Sprat T (1667) The history of the Royal Society of London. J. Martyn, London, UK

Stafford W (1978) Writing the Australian crawl: Views on the writer's vocation. University of Michigan Press, Ann Arbor.

Stapel Investigation (2012) Flawed science: The fraudulent research practices of social psychologist Diederik Stapel. https://www.tilburguniversity.edu/upload/3ff904d7-547b-40ae-85fe-bea38e05a34a_Final%20report%20Flawed%20Science.pdf

Steegmuller F (1947) Flaubert and Madame Bovary: A double portrait. New edition. Collins, London, UK

Steel P (2007) The nature of procrastination: A meta-analytic and theoretical review of quintessential self-regulatory failure. Psychological Bulletin 133:65–94

Steel P, König CJ (2006) Integrating theories of motivation. Academy of Management Review 31:889–913

Steinbeck J (1969) Journal of a novel: The East of Eden letters. (1972 edition) Pan, New York, NY.

Strunk W, Jr., White EB (1959) The elements of style. Macmillan, New York

Swales JM (1990) Genre analysis: English in academic and research settings. Cambridge University Press, Cambridge, UK

Swales JM (2004) Research genres: Explorations and applications. Cambridge University Press, Cambridge, UK

Swann WB, Hixon JG, Stein-Seroussi A, Gilbert DT (1990) The fleeting gleam of praise: Cognitive processes underlying behavioral reactions to self-relevant feedback. Journal of Personality and Social Psychology 59:17–26

Sword H (2012) Stylish academic writing. Harvard University Press, Cambridge, MA

Thaler R (1981) Some empirical evidence on dynamic inconsistency. Economics Letters 8:201–207

Thomas L (1979) The medusa and the snail: More notes of a biology watcher. Viking, New York, NY

Tufte ER (2001) The visual display of quantitative information. 2nd edition. Graphics Press, Cheshire, CT

Uzuner S (2008) Multilingual scholars' participation in core/global academic communities: A literature review. Journal of English for Academic Purposes 7:250–263

Van Doren M (1949) Nathaniel Hawthorne. W. Sloane Associates, New York, NY

Van Duzer C (2013) Sea monsters on medieval and Renaissance maps. British Library, London, UK

Van Valen LM (1973) A new evolutionary law. Evolutionary Theory 1:1–30

Vangeison G, Carr D, Federoff HJ, Rempe DA (2008) The good, the bad, and the cell type-specific roles of hypoxia inducible factor-1 alpha in neurons and astrocytes. Journal of Neuroscience 28:1988–1993

Vonnegut K, Jr. (1997) Timequake. Putnam, New York, NY

Wang Z, David P, Srivastava J, Powers S, Brady C, D'Angelo J, Moreland J (2012) Behavioral performance and visual attention in communication multitasking: A comparison between instant messaging and online voice chat. Computers in Human Behavior 28:968–975

Ware C (2012) Information visualization: Perception for design. 3rd edition. Morgan Kaufmann, Waltham, MA

West-Eberhard MJ (2014) Darwin's forgotten idea: The social essence of sexual selection. Neuroscience and Biobehavioral Reviews 46:501–508

Westfall RS (1980) Never at rest: A biography of Isaac Newton. Cambridge University Press, Cambridge, UK

Whitlock MC, McPeek MA, Rausher MD, Rieseberg LH, Moore AJ (2010) Data archiving. American Naturalist 175:145–146

Williams JM (1990) Style: Toward clarity and grace. University of Chicago Press, Chicago, IL

Wilson AT, Calvin M (1955) The photosynthetic cycle: CO_2-dependent transients. Journal of the American Chemical Society 77:5948–5957

Woods EC, Hastings AP, Turley NE, Heard SB, Agrawal AA (2012) Adaptive geographical clines in the growth and defense of a native plant. Ecological Monographs 82:149–168

Worm O (1655) Museum Wormianum. Elzevir, Leiden

Young HJ, Young TP (1992) Alternative outcomes of natural and experimental high pollen loads. Ecology 73:639–647

Zweig G (1964) An SU(3) model for strong interaction symmetry and its breaking. CERN Report Geneva TH. 401

URLs, and the web-based resources they point to, are notoriously ephemeral. Here, listed alphabetically by URL, are archived versions of all the web pages referred to in this book (via the Internet Archive Wayback Machine, http://archive.org/web/):

URL	Web page	Permanent URL (web archive)
bit.ly/1SaJBTy	"Cancer paper pulled due to 'identical text' from one published 6 days prior; author objects"	http://web.archive.org/web/2015 0612140339/http://retraction watch.com/2015/06/12/cancer -paper-pulled-due-to-identical -text-from-one-published-6-days -prior-author-objects/
cmap.ihmc.us	Cmap concept mapping software	http://web.archive.org/web/2015 0611182849/http://cmap.ihmc.us
jabref.sourceforge.net	Jabref reference management software	http://web.archive.org/web /20150422135218/http://jabref .sourceforge.net
scientistseessquirrel .wordpress.com	Scientist Sees Squirrel (my blog)	http://web.archive.org/web/2015 0422144106/https://scientistsees squirrel.wordpress.com
scientistseessquirrel .wordpress.com/2015 /04/24/dealing-with -the-defect-in-english/	"Dealing with the defect in English" (blog post)	http://web.archive.org/web/2015 0430124557/https://scientistsees squirrel.wordpress.com/2015/04 /24/dealing-with-the-defect-in -english/
supermag.jhuapl.edu	Supermag data archive	http://web.archive.org/web/2015 0422131515/http://supermag .jhuapl.edu/
suwtuesdays.word press.com	"Shut up and Write Tuesdays"	http://web.archive.org/web/2015 0422143342/https://suwtuesdays .wordpress.com/
www.750words.com	"750 words" writing tool	http://web.archive.org/web/20150 422134304/http://750words.com/
www.booknotes.org /FullPage.aspx?SID =60099–1	"Shelby Foote: Stars in Their Courses: The Gettysburg Campaign" (interview)	http://web.archive.org/web/201504 22133709/http://www.booknotes .org/FullPage.aspx?SID=60099–1

URL	Web page	Permanent URL (web archive)
www.ccdc.cam.ac.uk	Cambridge Crystallographic Data Center	http://web.archive.org/web/2015 0422131244/http://www.ccdc.cam .ac.uk/pages/Home.aspx
www.datadryad.org	Data Dryad archive	http://web.archive.org/web/2015 0430124335/http://www.data dryad.org
www.github.com	Github software development platform	http://web.archive.org/web/2015 0422135758/https://github.com/
www.hccfl.edu/media /181113/sigfigs.pdf	"A short guide to significant figures"	http://web.archive.org/web/2015 0422130722/http://www.hccfl.edu /media/181113/sigfigs.pdf
www.hemingwayapp .com	Hemingway complex-writing identifier	http://web.archive.org/web/2015 0422143716/http://www.heming wayapp.com/
www.icmje.org/ recommendations /browse/roles-and -responsibilities /defining-the-role-of -authors-and-contrib utors.html	"Defining the Role of Authors and Contributors"	http://web.archive.org/web/2015 0430123636/http://www.icmje.org /recommendations/browse/roles -and-responsibilities/defining-the -role-of-authors-and-contributors .html
www.indiana. edu/~wts/pamphlets /plagiarism.shtml	"Plagiarism: What It is and How to Recognize and Avoid It"	http://web.archive.org/web/201504 22125659/http://www.indiana .edu/~wts/pamphlets/plagiarism .shtml
www.mendeley.com	Mendeley reference management software	http://web.archive.org/web/2015 0422134645/https://www.mend eley.com/
www.murderyourdar lings.com/index.html	Murder Your Darlings (band) homepage	http://web.archive.org/web/201504 22134020/http://www.murder yourdarlings.com/index.html
www.ncbi.nlm.nih .gov/genbank/	Genbank archive	http://web.archive.org/web/2015 0422130829/http://www.ncbi.nlm .nih.gov/genbank/
www.orcid.org	ORCID identifier registry	http://web.archive.org/ web/20150422134515/http://orcid. org/
www.refman.com	Reference Manager reference managerment software	http://web.archive.org/web/2015 0422134735/http://www.refman .com

URL	Web page	Permanent URL (web archive)
www.refworks.com	Reference Works reference management software	http://web.archive.org/web /20150422135429/http://www .refworks.com
www.rescuetime.com	RescueTime software	http://web.archive.org/web /20150422134210/https://www .rescuetime.com/
www.siam.org /journals/plagiarism .php	"Authorial Integrity in Scientific Publication"	http://web.archive.org/web /20150422130251/http://www .siam.org/journals/plagiarism.php
www.stickk.com	Stickk goal registry	http://web.archive.org/web /20150422134411/http://www .stickk.com

INDEX

Pages references in italics denote illustrations or figures.

abbreviations, 178
abstracts, 77, 82–83, 200
acceptance (of submitted manuscripts), 219
acknowledgements, 126, 225, 277
acronyms, 178
Akerlof, George, 218
alchemists, 4
appendices, 127–30
archives, data, 129, 130–31
Asimov, Isaac, 11
assumed knowledge errors, 200
audience, 17–19, 57, 71–72, 176–77, 233–34;
    for grant proposals, 239; navigation costs
    for, 103, 127–28; self-revision and, 196–200
authority, 81, *92–93*, 132–33, 161, 165. *See
    also* credibility
authors: citations and, 133; coauthors as, 247–
    59; lists of, 81–82, 251–52; non-native En-
    glish writers as, 260–69; order of, 251–52
avoidance, 27–28

back matter sections, 77, 126–31
Bacon, Francis, 4–5, 247
Ballard, J. G., 43
beauty in writing, 280–84
binge/snack writing, 46–48
blogs, 244–45
book chapters, 234–36
Boyle, Robert, 5, *92*, 274–75
brevity, 182–90, 194–96, 201–2; appendices
    and, 128; in results sections, 69–70, 106–8,
    109, 186
bylines, 81–82

Calvin, M., 279
canonical structure. *See* structure
Cartland, Barbara, 11
central question statements. *See* summary
    statements
checklists of errors/writing problems, 180,
    202, 210
Cicero, 84

citations, 126–27, 132–37; self-revision and,
    202; tense usage in, 163–64
clarity in writing: audience and, 17–18; bene-
    fits of, 7–9; brevity and, 182; friendly/for-
    mal reviews and, 209; as goal, 6–7, 13;
    humor/pleasure and, 271, 273–86; linear
    order and, 65; sentences and, 160, 170;
    structure and, 75, 100; style and, 147; word
    choice and, 176
coauthors, 247–59, 267–68
color in figures, 114
comments papers, 144
commitment devices, 44–46
concept maps, 40, 63–65
conclusion sections, 69, 77, 124–25
condensed words, 178
content development. *See* story-finding
correspondence with editors/reviewers,
    219–21
cowriting, 253–58
craft of writing. *See* writing as craft
credibility, 91, 94–95, 165. *See also* authority
criticism, fear of, 28–29
cutting text. *See* omitted information

Darwin, Charles, 281
data archives, 129, 130–31
data compilations, 110–11
data reproductions, 110
decline, editorial (of submitted manuscripts),
    216
decline after review (of submitted manu-
    scripts), 218
decline without prejudice (of submitted man-
    uscripts), 216–17, 218
descriptive papers, 142–43
discipline, 43–46
discussion sections, 120–25; early writing
    and, 38–39; Introduction's relation to, 131;
    outlines and, 67; redundant content in, 186;
    scheduling time for, 47–48; self-revision
    and, 201; structure and, 77; tense usage in,
    163; writing backwards and, 69

disfavored reviewers, 215–16
dissertations, 240–41
distractions and interruptions, 28, 45, 48–49
drafts. *See* early writing
Dumas, Alexandre, 11

EAL writers, 260–69
early writing, 36, 38–41; momentum and, 50;
     self-revision and, 194, 205–6. *See also*
     story-finding
easing in, 40–41
editorial decisions: accept/decline continuum
     and, 216–19; correspondence on, 219–21;
     favored/disfavored reviewers and, 215–16;
     process of, 213–14
editors, 214, 219–21, 225
Eliot, T. S., 277–78
empty modifiers, 184
English as an Additional Language writers,
     260–69
environments, writing, 45, 48, 197–98

favored reviewers, 215–16
feedback. *See* review/revision process
feeling stuck. *See* writer's block
figures. *See* tables and figures
figure scales, 115–16
finding your story. *See* story-finding
Flaubert, Gustave, 12, 33
fMRI (functional magnetic resonance imag-
     ing), 23–24
Foote, Shelby, 22n
freewriting, 53
friendly reviews, 18–19, 26–27, 204–10,
     211–12
front matter sections, 77, 79–83
functional magnetic resonance imaging
     (fMRI), 23–24
Gell-Mann, Murray, 276
Google Scholar, 81, 126
Gordin, M. D., 260
grammar, 159–62
grant proposals, 96, 176, 238–40
Gross, A. G., 273n

Harmon, J. E., 273n
Hawthorne, Nathaniel, 7, 273
hedges, 120, 122–23, 184–85
Higgs, Peter, 218

Higham, N. J., 141
Hobbes, Thomas, 5
hourglass structure, 77, 84, 121
Hughes, Matt, 49–50
humor in writing, 80–81, 273–86
Hurlbert, S. H., 275, 279

impact factor, 133
IMRaD. *See* structure
initialisms, 178
inspiration, 37–38, 50
interruptions and distractions, 28, 45, 48–49
introduction sections, 84–88; citations in,
     133; discussion's relation to, 131; early writ-
     ing and, 38–39, 40; literature review in, 39,
     40; "methods-last" papers and, 138–39;
     outlines and, 67; redundant content in, 186;
     scheduling time for, 47–48; self-revision
     and, 200, 201; tense usage in, 163; writing
     backwards and, 69

jargon, 177–78
jokes. *See* humor in writing
Joyce, James, 276

King, Stephen, 7
Kolaczan, Chris, 94–95
Kroto, Harry W., 284

leaving things out. *See* omitted information
literary allusions, 276
literature review, 39, 40
logs, writing, 26–27
long words, 177–78, 184
loose threads strategy, 48–49

Martin, George R. R., 11
mathematical papers, 128, 134, 141–42,
     248
Maupertuis, Pierre-Louis de, *92*
Mermin, N. D., 282–83
metadiscourse, 185
metaphors, 277
"methods-last" papers, 138–39
methods papers, 163
methods sections, 89–98; early writing and,
     38, 40; historical development of, *92–93*;

"methods-last" papers and, 138–39; outlines and, 67; redundant content in, 186; scheduling time for, 48; self-revision and, 200; tense usage in, 162–63; writing backwards and, 69
modifiers, 184
Moffat, Chandra, 186, 189
momentum, 42–54, 200
monographs, 236–37
Montgomery, S. L., 144
multitasking abilities, 200

Nabokov, Vladimir, 281–82
name formats, 81–82
navigation costs, 103, 127–28
Newton, Isaac, 6
nominalizations, 178–79, 184
non-native English writers, 260–69
non-starting: intentional, 36–38; unintentional, 31–36
notes papers, 143
noun strings, 167, 170–71
numbers, 106–8

omitted information, 183, 194–96; in appendices, 128–30; assumed knowledge errors and, 200; in results sections, 69–70, 106–8, 109, 186
online publishing, 114; back matter and, 127; blogs and, 244–45; book chapters and, 235–36
online supplements, 76, 95, 106, 112, 127–30, 137; story-finding and, 62, 64, 65
Open Researcher and Contributor IDs (ORCID), 81–82
oral presentations, 237–38
ORCID (Open Researcher and Contributor IDs), 81–82
outlines: book chapters and, 235; early writing and, 40, 65–68; non-native English writers and, 262

padding, 184
paragraphs, 149–58
parallel constructions, 154, 157
parentheticals, 185
perfectionism, 28, 49–50
pers. comm., 135
pers. obs., 136

Pinker, Steven, 18, 172
plagiarism, 19–20, 96–97, 264, 266–67
poster presentations, 237–38
"power-of-attorney," publishing, 257–58
precision, 107
presentations, 237–38
procrastination, 31–36
pronoun antecedents, 156–57, 199
proposals. See grant proposals

Quiller-Couch, Arthur, 195
quotas, writing, 44

reader simulation, 196–200
reading, 17–21, 263
Red Queen Hypothesis, 277
redundant content, 106, 157, 186
references. See citations
relational devices, 154–56
repeatability, 93, 94–95
repetition, 154–55, 157
replication. See repeatability
responses to reviews, 207–9, 219, 222–30
results sections, 99–119; brevity in, 69–70, 106–8, 109, 186; "methods-last" papers and, 138–39; outlines and, 67; scheduling time for, 48; structure and, 77, 99–102; tables and figures in, 102–18; tense usage in, 163; writing backwards and, 69
review papers, 134, 139–41, 176
review/revision process, 18–19, 26–27, 191–92; early writing and, 36, 194; favored and disfavored reviewers in, 215–16; fear of criticism and, 28–29; formal reviews and, 211–21, 242–44; friendly reviews and, 18–19, 26–27, 204–10, 211–12; humor/pleasure and, 286; non-native English writers and, 267–69; response to reviews and, 207–9, 219, 222–30; scheduling time for, 48, 197–98; self-revision and, 193–203, 205–6
reviews: reading of, 207–9; writing of, 242–44
revisions, major/minor (of submitted manuscripts), 218–19
right-branching sentences, 172
Riley, James Whitcomb, 46
Rockman, Matt, 283
roundabout phrases, 184
Royal Society of London, 5

Sagi, I., 285
scales, figure, 115–16
schedules, 44–48, 197–98
schematics, 110
self-awareness, 22–29
sentences, 159–73; complexity in, 166–72; grammar and, 159–62; simplicity in, 162–66
serial friendly reviews, 209–10
significant digits, 107
snack writing, 46–48
software, 129
Sprat, Thomas, 5, 274
stack-loading sentences, 171–72
Stafford, William, 51
star formation example, 12–124, *58–60*, 66–68, 80, 90, 102
starting to write, 27–28, 30–41, 43; self-revision and, 200; summary statements and, 58–61
Steinbeck, John, 30
story-finding: audience and, 71–72; brevity and, 183; omitted information and, 69–70; online supplements and, *62, 64, 65*; planning tools and, 61–69; results sections and, 99; revisions in, 70; scheduling time for, 47–48; summary statements and, 57–61
stress positions, 169–70
structure, 74–78, 100; deviations from canonical, 138–45. *See also individual component parts (e.g., Introduction, etc.)*
Strunk Jr., William, 6–7
style, 147–48. *See also individual component parts (e.g., paragraphs, etc.)*
subhead outlines, 67
summary statements, 58–61, 66–67; grant proposals and, 239–40; titles as, 79–80
supplements, online. *See* online supplements
Sword, H., 273n

tables and figures: color in, 114; design of, 108–17; early writing and, 39, 40; humor/pleasure in, 278–79; redundant content in, 186; relating to text of, 117–18; in results sections, 102–18; scheduling time for, 48; story-finding and, 68–69
tautologous modifiers, 184
technical reports, 237
temporal motivation theory (TMT), 32–36, 46
tense, grammatical, 162–64

theory of mind, 196–97
theory papers, 141–42
theses, 240–42
Thomas, Lewis, 280–81
time of day, 45
titles, 79–81, 277
TMT. *See* temporal motivation theory
topic positions, 169–70
topic-sentences: internal word positions and, 169–70, 171–72; in outlines, 67–68; in paragraphs, 150–52; self-revision and, 199
transitional expressions, 155, 156, 199
translations, 261–62
troublesome words, 179–80

unfinished texts strategy, 48–49
unicorns, 278–79
unpubl. data, 136
unpubl. MS, 136

voice, grammatical, 164–66, 184
Vonnegut, Kurt, 49

Web of Science, 81, 126, 134
White, E. B., 6–7
Williams, J. M., 273n
Wilson, A. T., 279
witnessing: communal, 92; virtual, 92–93
word choice, 154, 174–81, 184–86
wordstacks, 62–63
writer's block, 28, 51–54
writing as craft, 11–14; historical development of, 3–6, *92–93*, 164–65, 177, 247–48, 260; humor/pleasure/beauty and, 273–86
writing as process, 13, 15–16; behavioral tricks and, 25–27; cowriting, 253–58; getting started, 27–28, 30–41, 43, 58–61; momentum and, 42–54; non-native English writers and, 261–64; review/revision as, 18–19, 26–27; self-awareness in, 22–29
writing backwards, 69
writing logs, 26–*27*
writing quotas, 44

Yechiam, E., 285

Zweig, George, 276